Springer Theses

Recognizing Outstanding Ph.D. Research

Aims and Scope

The series "Springer Theses" brings together a selection of the very best Ph.D. theses from around the world and across the physical sciences. Each volume has been selected for its scientific excellence by an internationally renowned institute or department. For greater accessibility to non-specialists, the published versions include an extended introduction, as well as a foreword by the student's supervisor explaining the special relevance of the work for the field. The series as a whole provides a valuable resource both for newcomers to the research fields, and for other scientists seeking detailed background information on special questions. Finally, it serves as an accredited documentation of the valuable contributions made by today's younger generation of scientists.

How to Apply

Theses prepared at internationally leading universities and institutes may be nominated for publication by supervisors with the endorsement of the head of department. Each such department is entitled to nominate one thesis per year. Enquiries and requests for a nomination form should be sent, together with the thesis, to Ms. Nathalie Berg (nathalie.berg@springer.com).

Eligibility Criteria

To qualify for consideration, theses should fulfil the following criteria:

- They must be written in good English.
- The topic should fall within the confines of Chemistry, Physics, Earth Sciences, Engineering and related interdisciplinary fields such as Materials, Nanoscience, Chemical Engineering, Complex Systems and Biophysics.
- The work reported in the thesis must represent a significant scientific advance.
- A maximum of 30% of the text may be a verbatim reproduction from the author's previous publications.
- They must have been examined and passed during the 12 months prior to nomination.
- The theses should have a clearly defined structure including an introduction accessible to new Ph.D. students and scientists not expert in the relevant field.
- After acceptance, theses should be supplemented by a "Supervisor's Foreword" outlining the wider scientific relevance of the work.

Indexed by zbMATH.

Naoya Iwamoto

Synthesis of Functional Characterization of Protein Domains for the Development of Mirror-Image Therapeutics

Doctoral Thesis accepted by
Kyoto University, Kyoto, Japan

Springer

Author
Dr. Naoya Iwamoto
Graduate School of Pharmaceutical Sciences
Kyoto University
Kyoto, Japan

Supervisor
Prof. Shinya Oishi
Laboratory of Medicinal Chemistry
Kyoto Pharmaceutical University
Kyoto, Japan

ISSN 2190-5053 ISSN 2190-5061 (electronic)
Springer Theses
ISBN 978-981-95-5365-5 ISBN 978-981-95-5366-2 (eBook)
https://doi.org/10.1007/978-981-95-5366-2

This Springer imprint is published by the registered company Springer Nature Singapore Pte Ltd.
The registered company address is: 152 Beach Road, #21-01/04 Gateway East, Singapore 189721, Singapore

Supervisor's Foreword

I am sincerely pleased that Dr. Naoya Iwamoto's doctoral thesis has been selected for inclusion in Springer Theses. Dr. Iwamoto joined our research group at Kyoto University as an undergraduate student in April 2018. His early work focused on applying the mirror-image protein of a target protein to a screening campaign for developing potential cancer therapeutics. In the spring of 2019, Dr. Iwamoto began his doctoral studies at the Graduate School of Pharmaceutical Sciences, Kyoto University, focusing on the synthesis and functional characterization of protein domains for the development of mirror-image protein therapeutics. During his doctoral research, Dr. Iwamoto successfully achieved the chemical synthesis of two target protein domains relevant to drug discovery: the extracellular domain of the T cell immunoreceptor with immunoglobulin and immunoreceptor tyrosine-based inhibitory motif domain (TIGIT), and domain III of human serum albumin (HSA). The bioactivities of these synthetic domains were validated by analyses of their interactions with respective binding partners. In addition, he engineered an artificial scaffold protein domain (monobody) to identify a synthetic variant with optimal structural properties and improved thermal stability. This enhanced molecular scaffold was subsequently applied to the design of mirror-image protein-based therapeutics. The synthesis of these complex protein domains presented substantial challenges because of the poor physicochemical properties of their intermediates and products. Nevertheless, through meticulous optimization and persistent refinement of synthetic strategies, Dr. Iwamoto established robust methods that enabled the successful preparation and functional analysis of these protein domains. These accomplishments were made possible through the generous cooperation and support of his collaborators: Prof. Toru Maruyama, Prof. Hiroshi Watanebe, Dr. Hitoshi Maeda (Kumamoto University) for the HSA project; Prof. Motohiro Nonaka (Kyoto University), and Prof. Hiroshi Murakami, and Dr. Gosuke Hayashi (Nagoya University) for the monobody project. Following the completion of his doctoral degree at Kyoto University, Dr. Iwamoto commenced his academic career as an assistant professor at Kyoto Pharmaceutical University in October 2024.

Dr. Iwamoto's thesis presents synthetic studies on several protein domains and their application in developing less-immunogenic mirror-image protein therapeutics.

He has published four articles derived from his thesis in journals of bioorganic chemistry and chemical biology. These contributions provide important benchmarks for the design and engineering of protein domains and scaffolds in drug discovery. I sincerely appreciate Dr. Iwamoto's dedication and perseverance in completing these ambitious and technically demanding projects.

Kyoto, Japan
October 2025

Prof. Shinya Oishi

List of Published Articles

Parts of this thesis have been published in the following journal articles:

Naoya Iwamoto, Jumpei Sasaki, Saya Ohno, Keisuke Aoki, Yusuke Usui, Shinsuke Inuki, Hiroaki Ohno, and Shinya Oishi, Synthetic studies on the extracellular domain of the T cell immunoreceptor with immunoglobulin and immunoreceptor tyrosine-based inhibitory motif domain (TIGIT) using Trt-K_{10} solubilizing tags. *Bioorg. Med. Chem.* **2024**, *99*, 117585.
Naoya Iwamoto, Takuma Kai, Shinsuke Inuki, Hiroaki Ohno, Hitoshi Maeda, Hiroshi Watanabe, Toru Maruyama, and Shinya Oishi, Mirror-image human serum albumin domain III as a tool for analyzing site II-dependent molecular recognition. *Bioconjugate Chem.* **2024**, *35*, 816–825.
Naoya Iwamoto, Yukino Sato, Asako Manabe, Shinsuke Inuki, Hiroaki Ohno, Motohiro Nonaka, and Shinya Oishi, Design and synthesis of monobody variants with low immunogenicity. *ACS Med. Chem. Lett.* **2023**, *14*, 1596–1601.
Naoya Iwamoto, Saya Ohno, Kensuke Nakamura, Toshinori Naito, Sayaka Miura, Shinsuke Inuki, Hiroaki Ohno, Gosuke Hayashi, Hiroshi Murakami, and Shinya Oishi, Design and evaluation of stable cysteine-modified monobody scaffolds for mirror-image synthesis. *Bioconjugate Chem.* **2025**, *36*, 1504–1515.

Acknowledgements

The author would like to express his deepest gratitude to Prof. Hiroaki Ohno (Graduate School of Pharmaceutical Sciences, Kyoto University) and Prof. Shinya Oishi (Kyoto Pharmaceutical University) for their kind directions, valuable discussions, and encouragement throughout this study.

The author also expresses his sincere and heartfelt gratitude to Prof. Shinsuke Inuki (Graduate School of Biomedical Sciences, Tokushima University), Dr. Norihito Arichi, and Dr. Hiroki Akiba (Graduate School of Pharmaceutical Sciences, Kyoto University) for their constructive discussions and sincere encouragement throughout this study.

The author is profoundly grateful to Mr. Jumpei Sasaki, Dr. Keisuke Aoki, Mr. Yusuke Usui (Graduate School of Pharmaceutical Sciences, Kyoto University), and Ms. Saya Ohno (Kyoto Pharmaceutical University) for their assistance with the synthesis of the extracellular domain of the T cell immunoreceptor with immunoglobulin and immunoreceptor tyrosine-based inhibitory motif domain; Prof. Toru Maruyama, Prof. Hiroshi Watanabe, Dr. Hitoshi Maeda, and Mr. Takuma Kai (Graduate School of Pharmaceutical Sciences, Kumamoto University) for their preparation of recombinant proteins of human serum albumin domain III; Prof. Motohiro Nonaka and Ms. Asako Manabe (Graduate School of Medicine, Kyoto University) for their immunogenicity assay of monobody proteins; Prof. Hiroshi Murakami, Dr. Gosuke Hayashi, Mr. Toshinori Naito, and Ms. Sayaka Miura (Graduate School of Engineering, Nagoya University) for their identification of Mb5-11; Ms. Yukino Sato, Ms. Saya Ohno, Mr. Kensuke Nakamura, Ms. Rina Takahashi, Ms. Ayaka Imai (Kyoto Pharmaceutical University) for their assistance with the synthesis of monobody variants.

The constructive advice from Prof. Shiroh Futaki and Prof. Hideaki Kakeya (Graduate School of Pharmaceutical Sciences, Kyoto University) is greatly appreciated.

The author is also grateful to all the colleagues in the Department of Bioorganic Medicinal Chemistry, Department of Biopharmaceutical Chemistry (Graduate School of Pharmaceutical Sciences, Kyoto University), and Laboratory of Medicinal Chemistry (Kyoto Pharmaceutical University) for their positive encouragement.

The author would like to show sincere gratitude to Fujita Memorial Scholarship from the late Professor Tetsuro Fujita and his bereaved family for financial support. The author also thanks the Japan Society for the Promotion of Science (JSPS) and Suntory Foundation for Life Sciences for financial support.

Contents

Chapter 1
Introduction

Abstract Protein domains are structural modules with distinct three-dimensional structures and biological functions. They have been widely used as important tools for drug discovery, including molecular probes for biochemical analysis and biopharmaceuticals themselves. Chemical synthesis enables the preparation of protein domains with high purity and homogeneity, as well as the modification with non-natural components. Especially, mirror-image protein domains, prepared by chemical synthesis using D-amino acids, should provide promising opportunities for exploring bioactive ingredients from non-natural chemical space. This thesis describes synthetic studies on several mirror-image protein domains of therapeutic importance, and their applications to developing non-immunogenic potential biopharmaceuticals.

Keywords Protein domain · Protein scaffold · Chemical protein synthesis · Mirror-image protein · Immunogenicity

Protein domains are structural modules found in naturally occurring proteins, which are characterized by distinct three-dimensional structures and biological functions [1]. Protein domains typically consist of 100–250 amino acids. Structural assignment of the protein sequences has suggested that approximately 80% of eukaryotic proteins consist of multiple domains [2]. These multi-domain proteins can be formed by gene duplication and recombination from common ancestral protein domains, along with sequence diversification via mutations, deletions, and insertions to generate novel molecular functions [2]. Although multi-domain proteins contribute to a great variety of physiological processes, limited numbers of protein domain families dominantly construct these multi-domain proteins in eukaryotes [3, 4].

In drug discovery, the use of protein domains has facilitated the precise analysis of the structures and functions of target proteins, as well as the development of biochemical assay systems. For example, the Src homology 2 (SH2) domain is a noncatalytic domain conserved among several cytosolic signaling proteins (Fig. 1.1). Pawson et al. have prepared fusion proteins of SH2 domains with bacterial TrpE to experimentally determine that SH2 domains play a pivotal role in the interaction of activated growth

N. Iwamoto, *Synthesis of Functional Characterization of Protein Domains for the Development of Mirror-Image Therapeutics*, Springer Theses,
https://doi.org/10.1007/978-981-95-5366-2_1

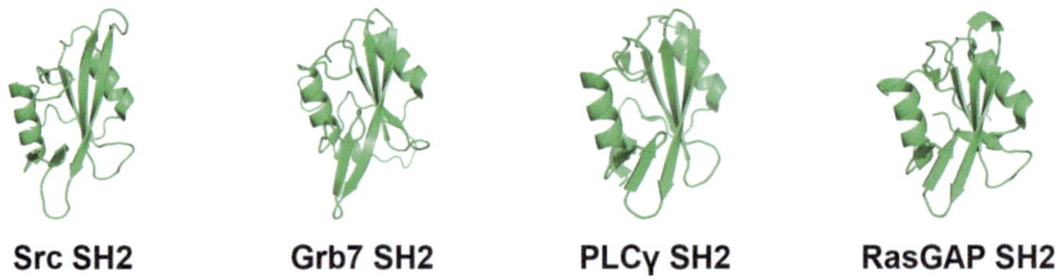

Fig. 1.1 Structures of SH2 domains in signaling proteins. [PDBID: 1NZL (Src SH2), 4WWQ (Grb7 SH2), 5TO4 (PLCγ SH2), 6PXB (RasGAP SH2)]

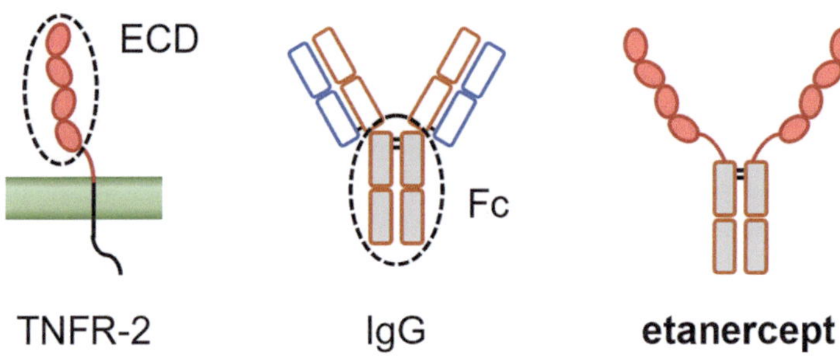

Fig. 1.2 Molecular design of etanercept

factor receptors with phosphotyrosine residues [5]. The molecular recognition of phosphorylated tyrosine by SH2 domains was investigated using mutant proteins [6]. These pioneering studies and subsequent research have led to an understanding of the distinct roles in cell signaling of the SH2 domains, which could be potential targets for cancer drug discovery [7]. An efficient screening of potential inhibitors of SH2 domains was first described by MacNeil et al. in 1997. High-throughput screening systems using fluorescence polarization were designed, in which the binding between a recombinant c-Src SH2 domain and a fluorescein-labeled phosphotyrosine peptide was detected [8].

Protein domains are also used as active pharmaceutical ingredients. One example of protein domains used as therapeutic modalities is fragment crystallizable (Fc)-fusion proteins, which consist of target-binding sequences and the Fc domains in human immunoglobulin G (IgG) [9]. The target-binding sequences can be derived from physiological ligands, functional sequences, and the extracellular domains (ECDs) of receptors to provide pharmacological effects. The Fc domains are beneficial for improving the pharmacokinetic profiles of fusion proteins through increasing the molecular sizes to avoid glomerular filtration, as well as for interacting with neonatal Fc receptors (FcRn) to enable recycling after endocytosis [10]. Etanercept is an Fc-fusion therapeutic protein (Fig. 1.2) for the treatment of rheumatoid arthritis [11]. In etanercept, the ECD of the human p75 tumor necrosis factor receptor (TNFR-2) functions as a decoy receptor for TNF to reduce the inflammatory response.

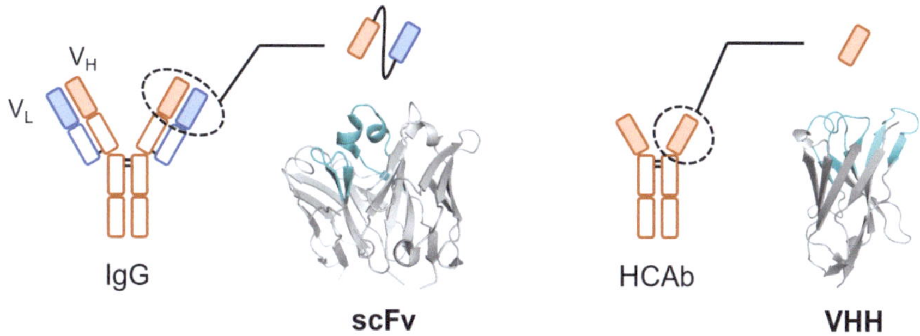

Fig. 1.3 Antibody fragments. Representative variable regions (CDRs) are highlighted in cyan. [PDBID: 2A9M (scFv), 7ZK1 (VHH)]

Antibody fragments are another promising type of protein domains for biopharmaceuticals (Fig. 1.3) [12], which are engineered from the variable domains of full-sized antibodies. The highly variable complementarity-determining regions (CDRs) on the β-sandwich immunoglobulin scaffolds of antibody fragments can mediate specific binding to target proteins [13, 14]. The single chain variable fragment (scFv) is a pioneering example of an antibody fragment, consisting of two variable domains, V_H in the heavy chain and V_L in the light chain of IgG, which are connected via a flexible peptide linker in a single chain (~ 28 kDa) [13]. For example, brolucizumab is an approved therapeutic scFv targeting vascular endothelial growth factor (VEGF) for the treatment of neovascular age-related macular degeneration [15]. A variable domain of the heavy chain of a heavy chain antibody (VHH; ~ 15 kDa) is also a successful example of antibody fragments used in various biological and therapeutic applications [14, 16]. Animals of the Camelidae family generate heavy chain antibodies (HCAb), which lack whole light chains and the first constant regions of the heavy chains in IgG. VHHs are the target-binding variable domains of HCAb and exhibit sufficient target binding and high stability even in a single domain. A representative example is caplacizumab, a bivalent VHH against von Willebrand factor, for the treatment of thrombotic thrombocytopenic purpura [17]. These antibody fragment-based therapeutics are advantageous in terms of efficient tissue penetration (as seen in brolucizumab) as well as having no Fc-mediated immune effector functions (as seen in caplacizumab) [18]. However, these antibody fragments may suffer from unfavorable pharmacokinetic properties, such as rapid clearance, because of the absence of Fc regions and their small architectures. To overcome these drawbacks, alternative approaches to prolong blood circulation are being developed [19].

Non-immunoglobulin protein binders have also been exploited as therapeutic modalities in drug discovery (Fig. 1.4) [20]. These artificial antibody-like protein binders are equipped with CDR-like variable regions on their parent protein domains. The highly stable scaffolds of the parent proteins can tolerate various modifications of the variable regions for specific and high-affinity binding to targets. There have been several protein domains employed as artificial antibody-like proteins, including the ankyrin repeat domain in the designed ankyrin repeat domain (DARPin) [21],

Fig. 1.4 Representative protein binders derived from non-immunoglobulin protein domains. Representative variable regions are highlighted in cyan. [PDBID: 5OOV (DARPin), 3UYO (monobody), 3MZW (afiibody), 4AFS (fynomer)]

fibronectin type III domain in monobodies [22], Z-domain of staphylococcal protein A in affibodies [23], and the SH3 domain of Fyn kinase in fynomers [24]. The particular properties of artificial antibody-like proteins derived from the parent protein domains can be appended to the therapeutic modalities of biopharmaceutics [18].

These protein-based drugs are generally produced using recombinant protein expression [25]. Recent advances in protein engineering have enabled the facile preparation of therapeutic proteins with high homogeneity and the correct folding using heterogenous expression systems [26]. Several therapeutic antibody fragments, including brolucizumab and caplacizumab, are currently produced by recombinant expression in *Escherichia coli*, although the presence of intramolecular disulfide bonds could possibly cause protein misfolding and aggregation [27].

Chemical protein synthesis is a versatile alternative to recombinant protein expression for the preparation of bioactive protein domains [28]. A major advantage of chemical protein synthesis is the high purity and homogeneity of the products because this method avoids the concomitant generation of partially degraded products and incomplete post-translational modifications (PTMs) [29]. Boutin et al. have conducted a comparative analysis of the purity of VHHs against anti-human epidermal growth factor 2 (HER2), which were prepared by chemical synthesis and recombinant expression [30]. The synthetic anti-HER2 VHH was highly homogenous, while the recombinant samples were somewhat heterogenous owing to reduced thiol groups or truncated C-terminal sequences. The homogenous properties of chemically synthesized materials are also favored for the accurate estimation of the impacts of specific modifications on the structures and/or functions [29]. For example, Tan et al. have comprehensively analyzed the effect of *O*-glycosylation in a fungus cellulase TrCel7A [31, 32]. TrCel7A consists of two protein domains: a catalytic domain to hydrolyze the cellulose of plant cell walls and a carbohydrate-binding module (CBM) to specifically recognize the polysaccharide structure. *O*-glycosylation in the CBM is involved in physiological stability and molecular recognition. The analysis, using several CBM variants with different states of *O*-glycosylation, revealed that *O*-glycosylation at Ser3 effectively improved the thermal stability and protease resistance of the CBM more than modifications at Thr1 or Ser14, (Fig. 1.5) [31], and naturally occurring α-*O*-mannose showed the most stabilizing effect [32].

Another advantage of chemical protein synthesis is the potential for modification with non-natural components, including D-amino acids [33]. Mirror-image proteins

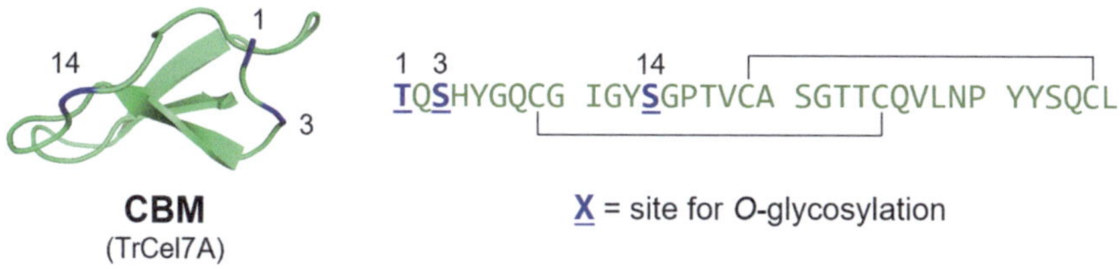

Fig. 1.5 Structure and amino acid sequence of CBM from TrCel7A. CBM variants were synthesized with *O*-glycosylation at the residues shown in blue. [PDBID: 1CBH]

(D-proteins) can be obtained by total chemical synthesis from D-amino acids. D-Proteins have enantiomeric structures to L-proteins and can have altered functions to the native L-proteins, providing attractive approaches for various research directions [34]. One of the major applications of D-proteins is racemic crystallography [35]. In 1993, Berg and Zawadzke reported that centrosymmetric protein crystals from a synthetic enantiomeric pair of rubredoxin proteins provided an electron density map with high quality [36]. A subsequent study by Kent et al. demonstrated that racemic crystallography was a complementary experimental technique for the crystal growth of crystallization-resistant proteins. Using this technique, the well-packed racemic crystals of snow flea anti-freeze protein (sfFiAFP) were obtained for crystallography [37].

D-Proteins have been used in drug discovery [38]. In 1996, Kim et al. proposed a screening strategy to identify D-peptide ligands for the native L-proteins of therapeutic targets, termed mirror-image phage display (MIPD) [39]. In their proof-of-concept study, a D-peptide ligand for the c-Src SH3 domain was identified (Fig. 1.6). In this technique, a synthetic D-protein of the Src SH3 domain (D-Src SH3) was employed for the phage display screening of a genetically encoded L-peptide library to identify D-Src SH3-binding L-peptides. Subsequently, a mirror-image form of the hit peptide was synthesized, which showed binding to the native L-Src SH3 domain. On the basis of this MIPD concept, several D-target protein domains have been chemically synthesized for phage display screening. These screening campaigns have led to the identification of potential D-peptide therapeutics against a p53-binding domain of MDM2 [40], an immunoglobulin-like variable (IgV) domain of programmed cell death protein 1 ligand (PD-L1) [41], an IgV domain of the T cell immunoreceptor with immunoglobulin and immunoreceptor tyrosine-based inhibitory motif domain (TIGIT) [42], and the cysteine-rich domain 2 (CRD2) of the human p55 TNF receptor (TNFR-1) [43]. In these examples, use of the relevant protein domains as synthetic and screening targets enabled the use of simple MIPD procedures to avoid the preparation of synthetically inaccessible full-length proteins [44].

Recently, the scope of the screening approach using D-target proteins has been extended to various chiral substances, including oligonucleotides [45], natural products [46], and protein binders (Fig. 1.7) [47–49]. Spiegelmers are aptamers composed entirely of L-RNA, which can be identified by screening using native D-oligonucleotide libraries [45]. The therapeutic L-oligonucleotides have resistance to

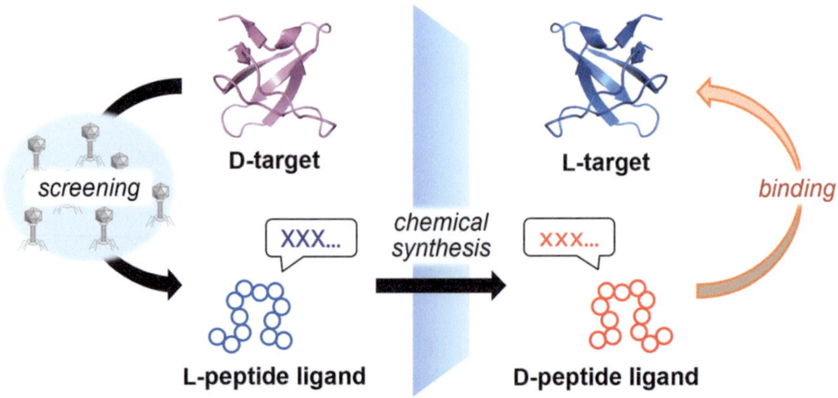

Fig. 1.6 Schematic representation of mirror-image phage display (MIPD) strategy. D-Target proteins are initially applied to phage display screening to identify L-peptide ligands. On the basis of amino acid sequences of the L-peptide ligands, D-peptide ligands are synthesized, which can bind to the native L-target protein

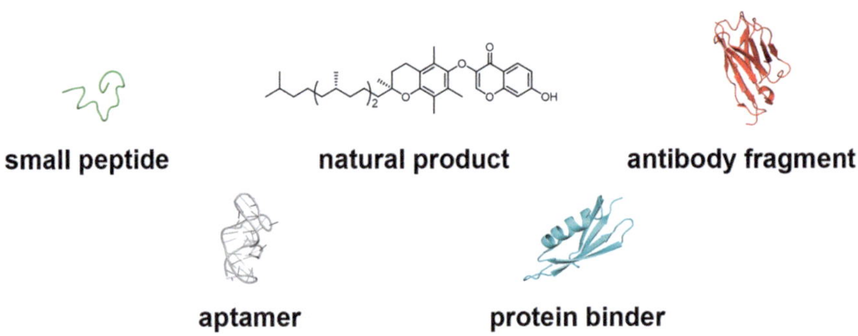

Fig. 1.7 Representative chiral substances employed for screening against D-target proteins

natural nuclease degradation as well as specific target binding capability. The author's group has previously reported the application of screening systems to chiral natural product libraries, leading to the identification of the MDM2 inhibitor *ent*-NP843, an α-tocopherol derivative with completely inverted stereochemical configurations [46]. Kent et al. have reported the discovery of several D-protein ligands for VEGF-A by MIPD [47–49]. The screening was conducted with libraries of artificial protein binders, which were derived from several non-immunoglobulin protein domains, namely the B1 domain of streptococcal protein G (56 a.a.), the GA domain from streptococcal protein G (53 a.a.), and the Z-domain from staphylococcal protein A (58 a.a.). These protein binders with a small molecular size were synthetically accessible and achieved specific target binding. The author's research group have also constructed a D-protein ligand, D-VHH [50, 51].

The advantage of this MIPD approach is the wide chemical space available for drug discovery as well as the favorable pharmaceutical properties of D-peptide and D-protein ligands. Kent et al. have demonstrated that proteases recognized the substrate peptides in a highly stereoselective manner, in which L-proteases selectively degraded L-peptide substrates and not D-peptide substrates [52]. In parallel, Berg et al. have demonstrated that a D-protein induced a reduced immune response compared with an L-protein upon in vivo administration [53]. These reduced immunogenic properties of the D-protein were consistent with several in vivo analyses of D-protein ligands [48–51]. Accordingly, D-peptides and D-proteins are promising drug modalities with high biological stability and low immunogenicity in vivo.

Recent advances in chemical protein synthesis technology has enabled the preparation of a variety of functional proteins with large molecular sizes and diverse structural modifications [54]. Current 9-fluorenylmethyloxycarbonyl (Fmoc)-based solid-phase peptide synthesis provides efficient access to longer peptides using rapid and automated procedures [55]. Additionally, several reliable chemistry techniques have been developed for the efficient ligation of two peptide segments at various positions in the sequences. These techniques are exemplified by native chemical ligation (NCL) [56], diselenide-selenoester ligation (DSL) [57], serine-threonine ligation (STL) [58], and KAHA ligation (Fig. 1.8) [59]. Other developed methods to handle insoluble synthetic intermediates and improve low refolding efficiency have also facilitated chemical protein synthesis [60]. Despite these advances in synthetic methodologies, the preparation of proteins of interest is often hampered by the nature of the sequences and unforeseeable molecular properties. Researchers may need to design and investigate the synthetic procedures for each target protein.

In the MIPD approach, functional evaluations of the synthetic proteins are a key process to validate the prepared D-proteins. Appropriate methods to assess the biological functions of D-proteins are often unavailable because D-proteins with their inverted stereochemistry cannot induce native biological responses. Specific probes and systems must be designed to assess the mirror-image functions. The development of practical methods for the synthesis and functional analysis of D-proteins would facilitate the application of D-protein domains as valuable tools for screening as well as the use of D-proteins as attractive therapeutic modalities in drug discovery.

In this thesis, the author describes the synthetic studies on three protein domains for mirror-image screening strategy. The author also reports the functional analyses of two target protein domains as well as investigation to improve the molecular properties of a synthetic mirror-image protein binder.

In Chapter 2, the author describes the chemical synthesis of the ECD of TIGIT [TIGIT(22–141); 120 residues], which is an attractive target molecule for developing potential immune checkpoint inhibitors.

In Chapter 3, the author describes the chemical synthesis of human serum albumin (HSA) domain III [HSA(380–585); 206 residues]. Functional analyses of site II-dependent molecular interaction using enantiomeric proteins are also reported.

In Chapter 4, the author describes the development of mirror-image monobody scaffold, an artificial protein binder derived from human fibronectin tenth type III

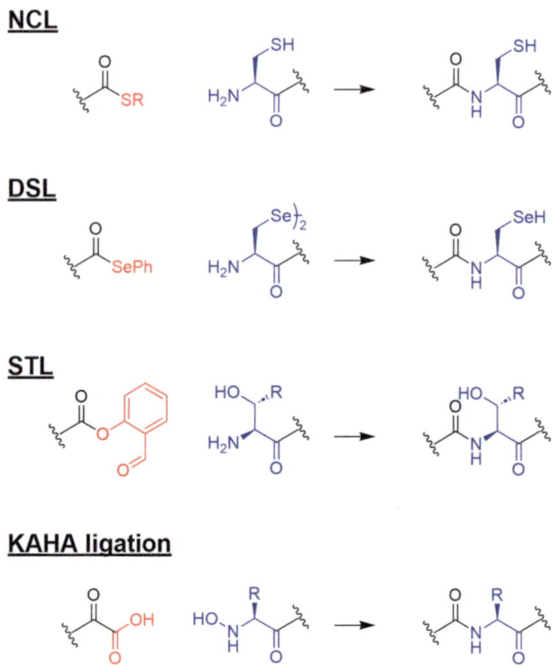

Fig. 1.8 Representative reactions of peptide ligation chemistries

domain. The structure–activity relationships of the scaffold architectures for the stabilization and functionalization are also reported.

References

1. Forslund K, Sonnhammer EL (2012) Evolution of protein domain architectures. Methods Mol Biol 856:187–216
2. Chothia C et al (2003) Evolution of the protein repertoire. Science 300:1701–1703
3. Koonin EV et al (2002) The structure of the protein universe and genome evolution. Nature 420:218–223
4. Yu L et al (2019) Grammar of protein domain architectures. Proc Natl Acad Sci USA 116:3636–3645
5. Anderson D et al (1990) Binding of SH2 domains of phospholipase C gamma 1, GAP, and Src to activated growth factor receptors. Science 250:979–982
6. Mayer BJ et al (1992) Point mutations in the abl SH2 domain coordinately impair phosphotyrosine binding in vitro and transforming activity in vivo. Mol Cell Biol 12:609–618
7. Chehayeb RJ, Boggon TJ (2020) SH2 Domain binding: Diverse FLVRs of partnership. Front Endocrinol 11:575220
8. Lynch BA et al (1997) A fluorescence polarization based Src-SH2 binding assay. Anal Biochem 247:77–82
9. Duivelshof BL et al (2021) Therapeutic Fc-fusion proteins: Current analytical strategies. J Sep Sci 44:35–62
10. Roopenian DC, Akilesh S (2007) FcRn: The neonatal Fc receptor comes of age. Nat Rev Immunol 7:715–725

11. Marotte H, Cimaz R (2014) Etanercept—TNF receptor and IgG1 Fc fusion protein: Is it different from other TNF blockers? Expert Opin Biol Ther 14:569–572
12. Harmansa S, Affolter M (2018) Protein binders and their applications in developmental biology. Development 145:148874
13. Ahmad ZA et al (2012) ScFv antibody: Principles and clinical application. Clin Dev Immunol 2012:980250
14. Muyldermans S (2013) Nanobodies: Natural single-domain antibodies. Annu Rev Biochem 82:775–797
15. Tadayoni R et al (2021) Brolucizumab: A newly developed anti-VEGF molecule for the treatment of neovascular age-related macular degeneration. Ophthalmologica 244:93–101
16. Beghein E, Gettemans J (2017) Nanobody technology: A versatile toolkit for microscopic imaging, protein–protein interaction analysis, and protein function exploration. Front Immunol 8:771
17. Hollifield AL et al (2020) Caplacizumab: An anti-von Willebrand factor antibody for the treatment of thrombotic thrombocytopenic purpura. Am J Health Syst Pharm 77:1201–1207
18. Vazquez-Lombardi R et al (2015) Challenges and opportunities for non-antibody scaffold drugs. Drug Discov Today 20:1271–1283
19. Kontermann RE (2011) Strategies for extended serum half-life of protein therapeutics. Curr Opin Biotech 22:868–876
20. Gebauer M, Skerra A (2020) Engineered protein scaffolds as next-generation therapeutics. Annu Rev Pharmacol Toxicol 60:391–415
21. Plückthun A (2015) Designed ankyrin repeat proteins (DARPins): Binding proteins for research, diagnostics, and therapy. Annu Rev Pharmacol Toxicol 55:489–511
22. Hantschel O et al (2020) Monobodies as enabling tools for structural and mechanistic biology. Curr Opin Struct Biol 60:167–174
23. Ståhl S et al (2017) Affibody molecules in biotechnological and medical applications. J Trends Biotechnol 35:691–712
24. Grabulovski D et al (2007) A novel, non-immunogenic Fyn SH3-derived binding protein with tumor vascular targeting properties. J Biol Chem 282:3196–3204
25. Assenberg R et al (2013) Advances in recombinant protein expression for use in pharmaceutical research. Curr Opin Struct Biol 23:393–402
26. Rosano GL, Ceccarelli EA (2014) Recombinant protein expression in Escherichia coli: Advances and challenges. Front Microbiol 5:172
27. Sandomenico A et al (2020) Evolution of Escherichia coli expression system in producing antibody recombinant fragments. Int J Mol Sci 21:6324
28. Tan Y et al (2020) Chemical protein synthesis: Advances, challenges, and outlooks. J Am Chem Soc 142:20288–20298
29. Chaffey PK et al (2018) Using chemical synthesis to study and apply protein glycosylation. Biochemistry 57:413–428
30. Hartmann L et al (2019) VHH characterization. Comparison of recombinant with chemically synthesized anti-HER2 VHH. Protein Sci 28:1865–1879
31. Chen L et al (2014) Specificity of O-glycosylation in enhancing the stability and cellulose binding affinity of Family 1 carbohydrate-binding modules. Proc Natl Acad Sci USA 111:7612–7617
32. Guan X et al (2015) Molecular-scale features that govern the effects of O-glycosylation on a carbohydrate-binding module. Chem Sci 6:7185–7189
33. Rohden F et al (2021) Through the looking glass: Milestones on the road towards mirroring life. Trends Biochem Sci 46:931–943
34. Harrison K et al (2023) Synthesis and applications of mirror-image proteins. Nat Rev Chem 7:383–404
35. Kent SBH (2018) Racemic & quasi-racemic protein crystallography enabled by chemical protein synthesis. Curr Opin Chem Biol 46:1–9
36. Zawadzke LE, Berg JM (1993) The structure of a centrosymmetric protein crystal. Proteins Struct Funct Genet 16:301–305

37. Pentelute BL et al (2008) X-ray structure of snow flea antifreeze protein determined by racemic crystallization of synthetic protein enantiomers. J Am Chem Soc 130:9695–9701
38. Zhao L, Lu W (2011) D-Peptide-based drug discovery aided by chemical protein synthesis. Isr J Chem 51:868–875
39. Schumacher TNM et al (1996) Identification of D-peptide ligands through mirror-image phage display. Science 271:1854–1857
40. Liu M et al (2010) A left-handed solution to peptide inhibition of the p53–MDM2 interaction. Angew Chem Int Ed 49:3649–3652
41. Chang HN et al (2015) Blocking of the PD-1/PD-L1 interaction by a D-peptide antagonist for cancer immunotherapy. Angew Chem Int Ed 54:11760–11764
42. Zhou X et al (2020) A novel D-peptide identified by mirror-image phage display blocks TIGIT/PVR for cancer immunotherapy. Angew Chem Int Ed 59:15114–15118
43. Lander AJ et al (2024) Deciphering the synthetic and refolding strategy of a cysteine-rich domain in the tumor necrosis factor receptor (TNF-R) for racemic crystallography analysis and D-peptide ligand discovery. ACS Bio Med Chem Au 4:68–76
44. Li Y et al (2021) Chemical synthesis of a full-length G-protein-coupled receptor β_2-adrenergic receptor with defined modification patterns at the C-terminus. J Am Chem Soc 143:17566–17576
45. Vater A, Klussmann S (2015) Turning mirror-image oligonucleotides into drugs: The evolution of Spiegelmer® therapeutics. Drug Discov Today 20:147–155
46. Noguchi T et al (2016) Screening of a virtual mirror-image library of natural products. Chem Commun 52:7653–7656
47. Mandal K et al (2012) Chemical synthesis and X-ray structure of a heterochiral D-protein antagonist plus vascular endothelial growth factor protein complex by racemic crystallography. Proc Natl Acad Sci USA 109:14779–14784
48. Uppalapati M et al (2016) A potent D-protein antagonist of VEGF-A is nonimmunogenic, metabolically stable, and longer-circulating in vivo. ACS Chem Biol 11:1058–1065
49. Marinec PS et al (2021) A non-immunogenic bivalent D-protein potently inhibits retinal vascularization and tumor growth. ACS Chem Biol 16:548–556
50. Aoki K et al (2023) Mirror-image single-domain antibody for a novel nonimmunogenic drug scaffold. Bioconjug Chem 34:2055–2065
51. Aoki K et al (2024) Engineering a low-immunogenic mirror-image VHH against vascular endothelial growth factor. ACS Chem Biol 19:1194–1205
52. Milton RC et al (1992) Total chemical synthesis of a D-enzyme: The enantiomers of HIV-1 protease show reciprocal chiral substrate specificity. Science 256:1445–1448
53. Dintzis HM et al (1993) A comparison of the immunogenicity of a pair of enantiomeric proteins. Proteins Struct Funct Genet 16:306–308
54. Agouridas V et al (2020) Chemical protein synthesis in medicinal chemistry. J Med Chem 63:15140–15152
55. Behrendt R et al (2016) Advances in Fmoc solid-phase peptide synthesis. J Pep Sci 22:4–27
56. Agouridas V et al (2019) Native chemical ligation and extended methods: Mechanisms, catalysis, scope, and limitations. Chem Rev 119:7328–7443
57. Bedding MJ et al (2022) Diselenide-selenoester ligation in the chemical synthesis of proteins. Methods Enzymol 662:363–399
58. Liu H, Li X (2018) Serine/threonine ligation: Origin, mechanistic aspects, and applications. Acc Chem Res 51:1643–1655
59. Bode JW (2017) Chemical protein synthesis with the α-ketoacid–hydroxylamine ligation. Acc Chem Res 50:2104–2115
60. Dong S et al (2024) Recent advances in chemical protein synthesis: Method developments and biological applications. Sci China Chem 67:1060–1096

Chapter 2
Synthetic Studies on the Extracellular Domain of the T Cell Immunoreceptor with Immunoglobulin and Immunoreceptor Tyrosine-Based Inhibitory Motif Domain (TIGIT) Using Trt-K10 Solubilizing Tags

Abstract The T cell immunoreceptor with immunoglobulin and immunoreceptor tyrosine-based inhibitory motif domain (TIGIT) is an inhibitory immunoreceptor expressed on lymphocytes that serves as a promising target for cancer immunotherapy. In this study, facile synthetic protocols to produce the extracellular domain of TIGIT were investigated for applications of TIGIT in mirror-image screening. During the synthesis via sequential native chemical ligations, the author encountered problems with significantly poor solubility of the ligated products. Introducing trityl-type solubilizing auxiliaries, which also functioned as temporary protecting groups for cysteine residues, facilitated a flexible order of ligations and efficient purification protocols. After refolding under appropriate conditions, the synthetic TIGIT showed a sufficient affinity toward its target ligand CD155.

Keywords T cell immunoreceptor with immunoglobulin and immunoreceptor tyrosine-based inhibitory motif domain · Immune checkpoint · Immunoglobulin variable domain · Chemical protein synthesis · Solubilizing tag

T cell immunoreceptor with immunoglobulin and immunoreceptor tyrosine-based inhibitory motif (ITIM) domain (TIGIT) is one of the polio virus receptor (PVR)/nectin family proteins [1]. TIGIT is expressed mainly on lymphocytes and interacts with other family protein ligands, such as CD155 and CD112 on tumor cells or dendritic cells in tumor microenvironments. These interactions upregulate negative regulatory signals of lymphocytes, leading to the suppression of immune responses. Thus, blocking the TIGIT-associated immune checkpoint can reactivate the immune response to facilitate cancer immunotherapy, which is similar to inhibitors of programmed death 1 and its ligand (PD-1/PD-L1) and cytotoxic T lymphocyte-associated antigen 4 (CTLA-4) [2]. Currently, several clinical trials targeting TIGIT are ongoing [3].

TIGIT consists of an extracellular immunoglobulin variable (IgV) domain, a transmembrane domain, an intracellular domain including ITIM and an immunoglobulin tail tyrosine (ITT)-like motif [1]. Because the extracellular IgV domain of TIGIT is responsible for the interaction with CD155 and other protein ligands,[4] several drug

N. Iwamoto, *Synthesis of Functional Characterization of Protein Domains for the Development of Mirror-Image Therapeutics*, Springer Theses,
https://doi.org/10.1007/978-981-95-5366-2_2

discovery efforts targeting the TIGIT-associated immune checkpoint have focused mainly on IgV-mediated interactions [5]. For example, a recent study revealed that a single chain variable fragment (scFv) targeting the IgV domain of TIGIT was identified from phage display screening, and its scFv-grafted monoclonal antibody showed potent anti-tumor activity [6]. Additionally, clinical anti-TIGIT antibodies recognize similar epitopes in the IgV domain to compete with CD155 binding [7]. Therefore, the extracellular IgV domain of TIGIT is a suitable target for screening against the TIGIT-associated checkpoint.

The author expected that the extracellular domain of TIGIT would be a promising target protein for mirror-image phage display screening strategy to explore potential anti-cancer agents. There have been a few reports on the chemical synthesis of an immunoglobulin variable domain of a TIGIT protein. Liu et al. previously established the synthetic process of the extracellular sequence of TIGIT (IgV^{TIGIT}; Thr24–Leu127) [8]. The mirror-image protein of IgV^{TIGIT} (D-IgV^{TIGIT}) was synthesized successfully and applied to mirror-image screening using phage display technology to identify D-IgV^{TIGIT}-binding sequences. The IgV^{TIGIT}-binding D-peptides suppressed tumor growth in a $CD8^+$ T cell-dependent manner. The removable backbone modification (RBM) strategy was employed for preparing the mirror-image protein of IgV^{TIGIT} to overcome difficulties in handling and isolating the peptide segments and the ligated products (Fig. 2.1) [9]. In this strategy, the poly-arginine moiety was conjugated to the peptide backbone, preventing the formation of secondary structures and unfavorable self-association to improve peptide solubility in aqueous solvent. Recently, chemical synthesis of IgV^{TIGIT} using backbone-installed split intein-assisted ligation (BISIAL) was also reported [10]. BISIAL enables peptide ligations at low concentrations, even in chaotropic solvents, which are promoted by the proximity effect of the backbone-installed split consensus-fast split intein. Alternatively, the author has attempted the straightforward synthesis of the extracellular domain of TIGIT ($TIGIT^{22–141}$; Met22–Pro141) using standard synthetic methods. In the attempt, although the author observed the formation of $TIGIT^{22–141}$, the highly insoluble properties of $TIGIT^{22–141}$ and its synthetic intermediates hampered reproducible and scalable synthesis. In this study, improved synthetic protocols for $TIGIT^{22–141}$ using trityl-type solubilizing auxiliary Trt-K_{10} [11] were investigated.

Initial Synthetic Attempts on the Extracellular Domain of TIGIT. $TIGIT^{22–141}$ contains three cysteine residues (Cys45, Cys69, Cys108), which can be used for native chemical ligations (NCLs) [12]. In the author's initial attempt, a designed middle peptide segment (Cys45–Phe107) was incompatible with subsequent ligations with the N-terminal segment (Met22–Gln44) or C-terminal segment (Cys108–Pro141) because of inefficient separation by preparative HPLC and low solubility of the products. Therefore, the author designed a synthetic strategy in which N-terminal (Met22–Ile68) and C-terminal (Cys69–Pro141) half-segments were initially constructed and then assembled at the Ile68–Cys69 junction to give full-length $TIGIT^{22–141}$ (Fig. 2.2). Four peptide segments **1**–**4** were designed and synthesized for this synthetic route. Peptide segment **1** (Met22–Gln44) was constructed on an *N*-acyl-*N'*-methyl-benzimidazolinone (MeNbz) linker [13]. According to a previous

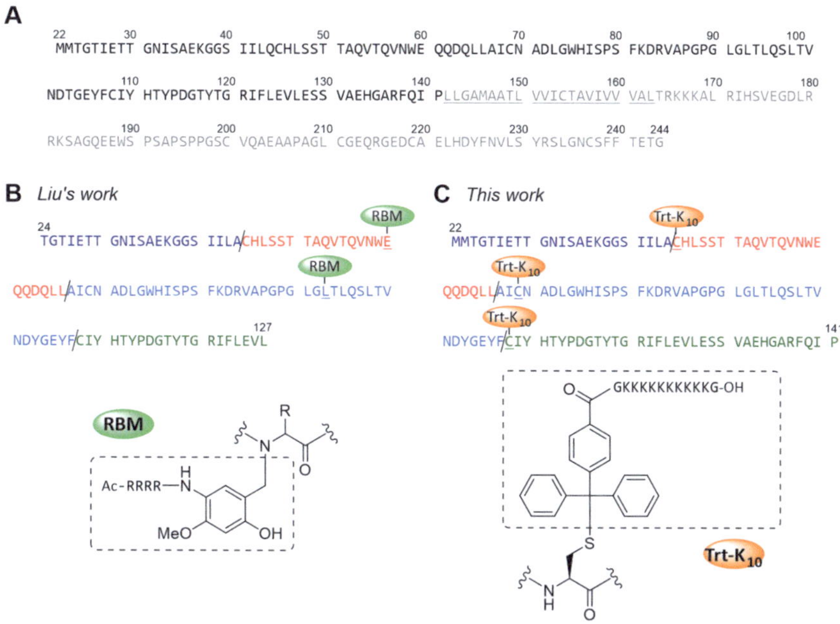

Fig. 2.1 The sequence of the TIGIT extracellular domain and the synthetic strategies. **a** Sequence of full-length TIGIT (Met22–Gly244). The sequence consists of the N-terminal extracellular domain (black), the transmembrane domain (gray, underlined) and the C-terminal intracellular domain (gray). **b** Design of peptide segments for synthesizing the immunoglobulin variable domain of TIGIT (Thr24–Leu127) by Liu et al. [8]. The four peptide segments are shown in different colors. Removable backbone modifications (RBM) were introduced at the residues underlined. **c** Design of peptide segments for the synthesis of the extracellular domain of TIGIT (Met22–Pro141) in this study. The four peptide segments are shown in different colors. Trt-K_{10} solubilizing tags were introduced at the cysteines underlined

report, replacing Gln44 with alanine had no influence on the binding with CD155 [8]. This substitution was favorable to avoid side reactions during NCLs [14]. Peptide segment **2** (Cys45–Ile68) was constructed on a diaminobenzoic acid (Dbz) linker [15], which can be converted into an active thioester. Peptide segment **3** (Cys69–Phe107) was constructed on the MeNbz linker at the C-terminus. The N-terminal Cys69 was protected as 1,3-thiazolidine-4-carboxylic acid (Thz), which was subsequently restored to cysteine. Poly-arginine tags were introduced at the C-terminus of peptide segments **1**–**3** to facilitate the purification of the peptide segments by improving solubility. Peptide segment **4** (Cys108–Pro141) was synthesized with a C-terminal poly-histidine tag, which can be used for immobilization and detection and improves solubility.

Initially, the author investigated the assembly of peptides **1** and **2**. 1,2,4-Triazole-mediated NCL [16] between **1** and **2** proceeded efficiently to form peptide **5**; however, peptide **5** was poorly eluted from a preparative HPLC column, resulting in failure to isolate **5** with sufficient purity and yield (*ca.* 13%). To facilitate the purification of

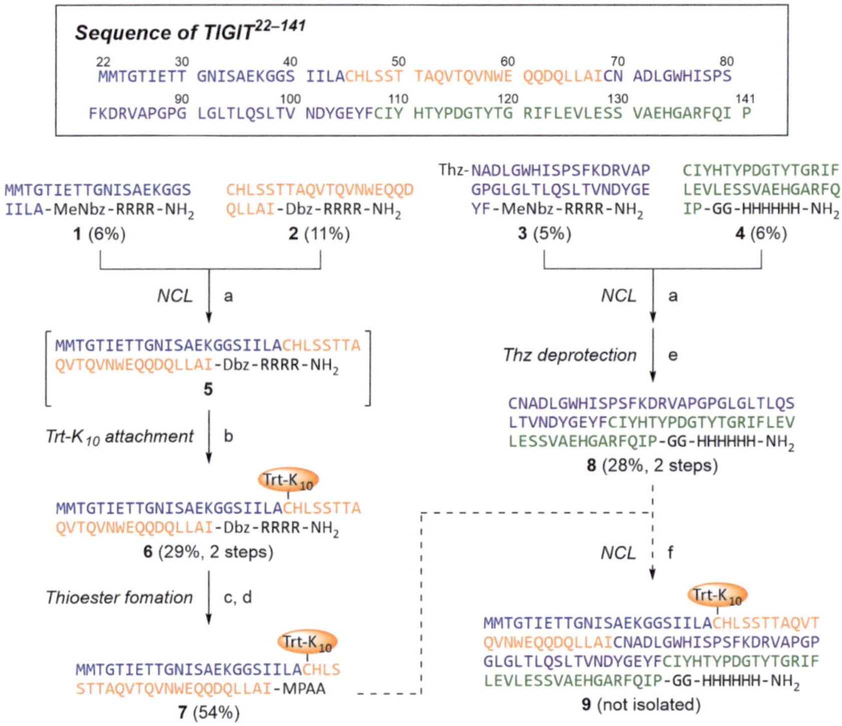

Fig. 2.2 Original synthetic plan of TIGIT$^{22-141}$ via the Ile68–Cys69 junction. *Reagents and conditions:* **a** 1,2,4-triazole, TCEP, 6 M guanidine, 100 mM phosphate buffer (pH 7.0); **b** Trt(OH)-K_{10}, TFA; **c** $NaNO_2$, 6 M guanidine, 100 mM phosphate buffer (pH 3.0); **d** MPAA, TCEP, 6 M guanidine, 200 mM phosphate buffer (pH 6.5); **e** methoxyamine; **f** MPAA, TCEP, 6 M guanidine, 200 mM phosphate buffer (pH 7.0)

5, the author focused on using the solubilizing auxiliary Trt-K_{10} (Fig. 2.1) [11]. Trt-K_{10} can enhance the solubility of conjugated peptides significantly by introducing a highly hydrophilic poly-lysine sequence at the side chain(s) of cysteine residue(s) through the detachable trityl linker. The reaction mixture of NCL between peptides **1** and **2** was treated with Trt(OH)-K_{10} in the presence of TFA to form peptide **6**, in which Cys45 was modified with Trt-K_{10}. In contrast to peptide **5**, peptide **6** was eluted efficiently from the preparative HPLC column and isolated with a satisfactory yield (29%). Trt-K_{10} was stable under subsequent conditions for $NaNO_2$-mediated activation of the Dbz linker [17] in **6** to provide an MPAA-thioester form of N-terminal half-part **7**. Separately, C-terminal half-part **8** was synthesized by 1,2,4-triazole-mediated NCL between peptides **3** and **4** followed by one-pot deprotection of the Thz group with methoxyamine. The author then investigated the ligation between peptides **7** and **8** at the sterically-hindered Ile–Cys junction. The NCL between **7** and **8** in the presence of MPAA proceeded to some extent to form the expected product **9**, which was detected by LC–MS analysis. However, after a long reaction time [18], the production of a significant quantity of unidentified byproducts was observed, which hampered the isolation of **9** with sufficient purity by this synthetic process, presumably because of the relatively low stability of Trt-K_{10} under the reaction conditions used.

Improved Synthesis of the Extracellular Domain of TIGIT. The author envisioned that the synthesis of full-length $TIGIT^{22-141}$ may be achieved by conjugation at Leu66–Ala67 instead of Ile68–Cys69 as the initial attempt. Because the Leu thioester at the peptide C-terminus is less sterically-hindered [18], NCL at the Leu–Cys junction would be completed before the Trt-K_{10} moieties are damaged. The author also focused on a previous report showing that Trt-K_{10} can function as a protecting group for cysteine residues during a metal-free desulfurization process [19]. In the newly designed synthetic strategy (Fig. 2.3), Thz at Ala67 can be converted into cysteine by methoxyamine-mediated deprotection, which is used for the NCL and then restored to alanine by metal-free desulfurization [20]. During this desulfurization, three native cysteines (Cys45, Cys69, Cys108) in the sequence were protected by Trt-K_{10}. In peptide ligations for chemical protein syntheses, Xaa–Ala junctions are generally constructed in earlier steps so the desulfurization reaction does not damage native cysteine residues. In the current synthetic strategy, using Trt-K_{10} as a protecting group enabled a flexible order of ligations to achieve a more convergent and efficient synthesis of the target protein.

Peptide segments **10** (Cys45–Leu66) and **11** (Ala67–Phe107) were newly designed and synthesized by the same protocol used for the synthesis of peptides **2** and **3**, respectively, to investigate this revised synthetic process of $TIGIT^{22-141}$. 1,2,4-Triazole-mediated NCL between peptides **1** and **10** followed by modification of **12** with Trt-K_{10} provided peptide **13**, which was converted into MPAA ester **14**. In the presence of MPAA, NCL between peptides **11** and **4** gave peptide **15**. Two sulfhydryl groups at Cys69 and Cys108 in peptide **15** were treated with Trt(OH)-K_{10} in the presence of TFA to give peptide **16** with Trt-K_{10} modifications. Although the one-pot NCL procedure and subsequent Trt-K_{10} attachment were feasible, the author isolated peptides **15** and **16** at each step to remove byproducts completely. The Thz group of **16** was treated with methoxyamine to provide cysteine peptide **17,** and then the NCL conditions between peptides **14** and **17** were investigated. As expected, ligation at the Leu66–Cys67 junction proceeded smoothly to give **18**, in which all native cysteines were protected by Trt-K_{10} and the temporary cysteine (Cys67) was unprotected. Peptide **18** was subjected to metal-free desulfurization conditions to successfully yield Trt-K_{10}-protected $TIGIT^{22-141}$ (**19**). In the final step, TFA/TIS treatment of **19** for detaching the Trt-K_{10} tag afforded full-length $TIGIT^{22-141}$ (**20**). Peptide **20** was purified by size-exclusion chromatography with a chaotropic solvent because of poor solubility in the HPLC solvent. The current revised strategy successfully provided $TIGIT^{22-141}$ from four peptide segments, which were synthesized by the standard Fmoc-SPPS without using backbone modification(s). The overall yield of the synthetic $TIGIT^{22-141}$ (1.4%, 6 steps from peptide **4**) was comparable to that in the previous report [10].

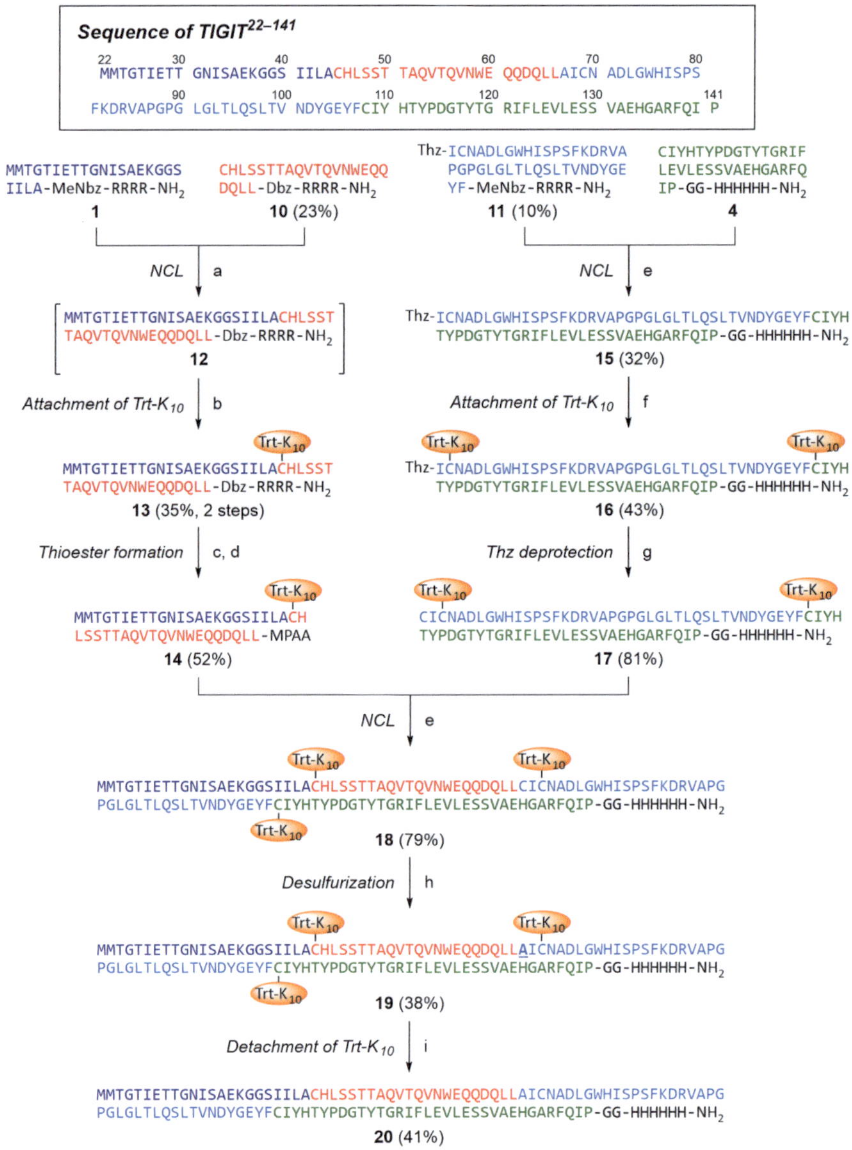

Fig. 2.3 Synthesis of TIGIT$^{22–141}$ via the NCL between the Leu66–Ala67 junction. *Reagents and conditions:* **a** 1,2,4-triazole, TCEP, 6 M guanidine, 100 mM phosphate buffer (pH 7.0); **b** Trt(OH)-K_{10}, TFA; **c** $NaNO_2$, 6 M guanidine, 100 mM phosphate buffer (pH 3.0); **d** MPAA, TCEP, 6 M guanidine, 200 mM phosphate buffer (pH 6.5); **e** MPAA, TCEP, 6 M guanidine, 200 mM phosphate buffer (pH 7.0); **f** Trt(OH)-K_{10}, TFA, 6 M guanidine, 200 mM phosphate buffer (pH 7.0); **g** methoxyamine; **h** VA-044, TCEP, MESNa, 6 M guanidine, 100 mM phosphate buffer (pH 6.5); **i** TIS, TFA

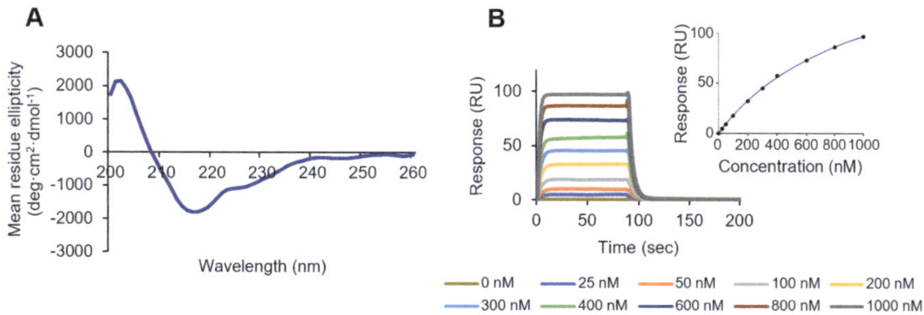

Fig. 2.4 Characterization of TIGIT$^{22-141}$. **a** CD spectrum of refolded TIGIT$^{22-141}$. **b** SPR sensorgrams of the association, steady-state binding, and dissociation interactions of TIGIT$^{22-141}$ and CD155. Various concentrations of TIGIT$^{22-141}$ were injected over immobilized CD155 ($\approx$ 1000 RU). The K_D value was determined from a saturation curve generated from triplicate experiments ($K_D = 0.95 \pm 0.02$ μM)

Functional Evaluation of the Synthetic TIGIT$^{22-141}$. With full-length TIGIT$^{22-141}$ in hand, the author investigated the refolding procedures according to the previous report with some modifications [21]. Denatured TIGIT$^{22-141}$ was diluted in HEPES buffer containing 400 mM arginine, 3 mM EDTA, 5 mM reduced glutathione and 0.5 mM oxidized glutathione, to provide refolded TIGIT$^{22-141}$. The refolded protein showed a circular dichroism (CD) spectrum with a minimum around 216 nm, suggesting the presence of the reported β-sheet-rich structure (Fig. 2.4a) [8]. The bioactivity of refolded TIGIT$^{22-141}$ was confirmed by surface plasmon resonance (SPR) analysis of the interaction with the extracellular domain of human CD155 (Fig. 2.4b). The determined K_D between synthetic TIGIT$^{22-141}$ and CD155 was 0.95 ± 0.02 μM, which is comparable to previous reports (K_D = 0.36–3.2 μM) [21, 22]. This is the first report describing the bioactivity of synthetic TIGIT toward CD155. These results support that a functional protein of the extracellular IgV domain of TIGIT was successfully constructed by chemical synthesis.

In summary, the author has successfully achieved the chemical synthesis of the extracellular domain of TIGIT. The author used the trityl-type solubilizing auxiliary Trt-K_{10} to improve the solubility of the synthetic intermediates. The author also demonstrated the potential of Trt-K_{10} as a protecting group of cysteine residues, which enabled the design of a rational synthetic route, including late-stage peptide assembly at the Xaa–Ala junction. The resulting TIGIT$^{22-141}$ showed satisfactory binding to the target CD155, which indicated that synthetic TIGIT$^{22-141}$ refolded to the bioactive structure.

2.1 Experimental Section

General Procedures of Peptide Synthesis. All reagents and solvents were purchased from Watanabe Chemical Industries, Ltd. (Hiroshima, Japan), Kokusan Chemical Industries, Ltd. (Kanagawa, Japan), Sigma-Aldrich JAPAN (Tokyo, Japan), Wako Pure Chemical Industries, Ltd. (Osaka, Japan), Tokyo Chemical Industry Co., Ltd. (Tokyo, Japan) or Nacalai Tesque, Inc. (Kyoto, Japan). For analytical HPLC, a Cosmosil 5C18-AR300 column (Nacalai Tesque, 4.6 × 250 mm) or a Cosmosil 5C4-AR300 column (Nacalai Tesque, 4.6 × 150 mm) was employed with a linear gradient of CH_3CN containing 0.1% (v/v) TFA or 0.05% (v/v) TFA at a flow rate of 1 mL/min. The products were detected by UV absorbance at 220 nm. For preparative HPLC, a Cosmosil 5C18-AR300 preparative column (Nacalai Tesque, 20 × 250 mm, flow rate 8 mL/min), a Cosmosil Protein-R preparative column (Nacalai Tesque, 20 × 150 mm, flow rate 8 mL/min) or a Cosmosil 5C4-AR300 preparative column (Nacalai Tesque, 20 × 150 mm, flow rate 8 mL/min) was employed with a linear gradient of CH_3CN containing 0.1% (v/v) TFA or 0.05% (v/v) TFA at a flow rate of 8 mL/min. All peptides were characterized by ESI–MS [micromass ZQ (Waters), LCMS-2020 (Shimadzu)].

Fmoc-SPPS. Standard Fmoc-SPPS was performed using a PSSM-8 automatic peptide synthesizer (Shimadzu, Kyoto, Japan) or a CYCROSS automatic peptide synthesizer (DFC, Kyoto, Japan). The following side chain protected amino acids were employed: Arg(Pbf), Asn(Trt), Asp(O*t*-Bu), Cys(Trt), Gln(Trt), Glu(O*t*-Bu), His(Trt), Lys(Boc), Ser(*t*-Bu), Thr(*t*-Bu), Trp(Boc), Tyr(*t*-Bu). Fmoc-protected amino acids were coupled using OxymaPure (5 equiv)/DIC (5 equiv) or HBTU (5 equiv)/HOBt·H_2O (5 equiv)/DIEA (10 equiv) in DMF for 60 min twice and Fmoc-deprotection was performed by 20% piperidine in DMF for 4 min twice or 5 min twice.

Synthesis of Peptide Segments with a MeNbz Moiety at the C-terminus: Synthesis of TIGIT$^{22–44}$ (1). Poly-arginine tag was manually loaded on Fmoc-NH SAL resin (0.57 mmol/g, 0.14 g, 0.080 mmol) by Fmoc-Arg(Pbf)-OH (0.13 g, 0.20 mmol), OxymaPure (28 mg, 0.20 mmol) and DIC (31 μL, 0.20 mmol) in DMF for 2 h each step. MeDbz linker was then loaded on the resin by Fmoc-MeDbz-OH (78 mg, 0.20 mmol), HATU (76 mg, 0.20 mmol) and DIEA (70 μL, 0.40 mmol) in DMF for 3 h. The following Ala44 was coupled by Fmoc-Ala-OH (0.13 g, 0.40 mmol), HATU (0.15 g, 0.40 mmol) and DIEA (0.14 mL, 0.80 mmol) in DMF for 2 h. The remaining peptide sequence was constructed by a standard protocol of Fmoc-SPPS. Boc-Met-OH was employed for coupling of the N-terminal Met22. After the solid-phase peptide synthesis, a solution of *p*-nitrophenyl chloroformate in CH_2Cl_2 (90 mg/mL, 2.0 mL) was added to the protected peptide resin, and the reaction was continued for 2 h at room temperature. Subsequently, the resin was treated with 1 M DIEA in DMF (2.0 mL) for 2 h at 37 °C twice. The resin was then treated with TFA/H_2O/*m*-cresol/thioanisole/EDT (80:5:5:5:5) for 2 h at room temperature. After removal of the resin by filtration, the filtrate was poured into ice-cold dry Et_2O.

The precipitate was purified by preparative HPLC (Cosmosil 5C4-AR300 column, a linear gradient of 15–45% CH_3CN in H_2O containing 0.1% (v/v) TFA over 90 min) to afford **1** (18.0 mg, 6% from resin). MS(ESI) *m/z*, calcd for $C_{129}H_{222}N_{44}O_{40}S_2$: 3091.61; observed: $[M + 2H]^{2+}$ *m/z* = 1547.18, $[M + 3H]^{3+}$ *m/z* = 1031.77, $[M + 4H]^{4+}$ *m/z* = 773.97.

TIGIT$^{45–68}$ (2). By the standard protocol for peptide synthesis, peptide **2** was synthesized (93.2 mg, 11% yield) from Fmoc-NH SAL resin (0.43 g, 0.20 mmol). MS(ESI) *m/z*, calcd for $C_{147}H_{240}N_{52}O_{44}S$: 3469.79; observed: $[M + 3H]^{3+}$ *m/z* = 1158.15, $[M + 4H]^{4+}$ *m/z* = 868.80.

[Thz^{69}]-TIGIT$^{69–107}$ (3). By the same protocol for the synthesis of peptide **1**, peptide **3** was synthesized (30.3 mg, 5% yield) from Fmoc-NH SAL resin (0.21 g, 0.10 mmol). MS(ESI) *m/z*, calcd for $C_{221}H_{338}N_{68}O_{63}S$: 4984.51; observed: $[M + 3H]^{3+}$ *m/z* = 1663.40, $[M + 4H]^{4+}$ *m/z* = 1247.80, $[M + 5H]^{5+}$ *m/z* = 998.40.

TIGIT$^{108–141}$ (4). By the standard protocol for peptide synthesis, peptide **4** was synthesized (28.3 mg, 6% yield) from Fmoc-NH SAL PEG resin (0.48 g, 0.080 mmol). MS(ESI) *m/z*, calcd for $C_{216}H_{310}N_{66}O_{59}S$: 4804.30; observed: $[M + 4H]^{4+}$ *m/z* = 1203.05, $[M + 5H]^{5+}$ *m/z* = 962.29, $[M + 6H]^{6+}$ *m/z* = 802.21.

TIGIT$^{22–68}$ (5). Peptides **1** (2.0 mg) and **2** (1.4 mg) were dissolved in 1,2,4-triazole ligation buffer [2.5 M 1,2,4-triazole, 30 mM TCEP, 6 M guanidine, 100 mM phosphate (pH 7.0); 0.33 mL], and the reaction mixture was incubated for 1 h at 37 °C. The mixture was purified by preparative HPLC (Cosmosil 5C4-AR300 column, a linear gradient of 22–52% CH_3CN in H_2O containing 0.05% (v/v) TFA over 90 min) to provide **5** including small impurities (0.3 mg, *ca.* 13% yield). MS(ESI) *m/z*, calcd for $C_{243}H_{405}N_{77}O_{78}S_3$: 5745.93; observed: $[M + 3H]^{3+}$ *m/z* = 1917.15, $[M + 4H]^{4+}$ *m/z* = 1438.05, $[M + 5H]^{5+}$ *m/z* = 1150.55, $[M + 6H]^{6+}$ *m/z* = 958.90.

[Cys(Trt-K$_{10}$)^{45}]-TIGIT$^{22–68}$ (6). Peptides **1** (15.8 mg) and **2** (11.1 mg) were dissolved in 1,2,4-triazole ligation buffer (2.6 mL), and the reaction mixture was incubated for 1 h at 37 °C. Then, Trt(OH)-K_{10} (31.2 mg) and TFA (2.6 mL) were added, and the reaction mixture was incubated for 1 h at room temperature. The mixture was purified by preparative HPLC (Cosmosil 5C18-AR300 column, a linear gradient of 22–52% CH_3CN in H_2O containing 0.05% (v/v) TFA over 90 min) to provide **6** (7.2 mg, 29% yield). MS(ESI) *m/z*, calcd for $C_{327}H_{545}N_{99}O_{92}S_3$: 7427.02; observed: $[M + 4H]^{4+}$ *m/z* = 1858.80, $[M + 5H]^{5+}$ *m/z* = 1487.15, $[M + 6H]^{6+}$ *m/z* = 1239.40, $[M + 7H]^{7+}$ *m/z* = 1062.45.

[Cys(Trt-K$_{10}$)^{45}]-TIGIT$^{22–68}$ MPAA ester (7). A $NaNO_2$ solution [200 mM $NaNO_2$, 6 M guanidine (pH ~ 3.0); 96 μL] was added to a solution of **6** (9.1 mg) in activation buffer [6 M guanidine, 100 mM phosphate buffer (pH 3.0); 0.96 mL] at − 20 °C, and the reaction was continued for 30 min. An MPAA solution [200 mM MPAA, 100 mM TCEP, 6 M guanidine, 200 mM phosphate buffer (pH 6.5); 0.32 mL] was then added to the reaction, and the reaction mixture was stirred for 10 min at − 20 °C. The mixture was purified by preparative HPLC (Cosmosil Protein-R column, a linear gradient of 28–58% CH_3CN in H_2O containing 0.05% (v/v) TFA over

90 min) to provide **7** (4.3 mg, 54% yield). MS(ESI) *m/z*, calcd for $C_{304}H_{496}N_{80}O_{89}S_4$: 6819.56; observed: $[M + 4H]^{4+}$ *m/z* = 1706.90, $[M + 5H]^{5+}$ *m/z* = 1365.65.

TIGIT$^{69–141}$ (8). Peptides **3** (4.5 mg) and **4** (3.0 mg) were dissolved in 1,2,4-triazole ligation buffer (0.50 mL), and the reaction mixture was incubated for 1 h at 37 °C. Then, 2 M $MeONH_2$ (59 μL) was added, and the reaction mixture was incubated for 2 h at 37 °C. The mixture was purified by preparative HPLC (Cosmosil 5C18-AR300 column, a linear gradient of 30–60% CH_3CN in H_2O containing 0.05% (v/v) TFA over 90 min) to provide **8** (1.5 mg, 28% yield). MS(ESI) *m/z*, calcd for $C_{403}H_{591}N_{115}O_{116}S_2$: 8961.33; observed: $[M + 5H]^{5+}$ *m/z* = 1794.25, $[M + 6H]^{6+}$ *m/z* = 1495.40, $[M + 7H]^{7+}$ *m/z* = 1281.90, $[M + 8H]^{8+}$ *m/z* = 1121.70.

TIGIT$^{45–66}$ (10). By the standard protocol for peptide synthesis, peptide **10** was synthesized (37 mg, 23% yield) from Fmoc-NH SAL resin (0.14 g, 0.080 mmol). MS(ESI) *m/z*, calcd for $C_{138}H_{224}N_{50}O_{42}S$: 3285.66; observed: $[M + 2H]^{2+}$ *m/z* = 1643.73, $[M + 3H]^{3+}$ *m/z* = 1096.27, $[M + 4H]^{4+}$ *m/z* = 822.30.

[Thz^{67}]-TIGIT$^{67–107}$ (11). By the same protocol for the synthesis of peptide **1**, peptide **11** was synthesized (48.7 mg, 10% yield) from Fmoc-NH SAL resin (0.14 g, 0.080 mmol). MS(ESI) *m/z*, calcd for $C_{230}H_{354}N_{70}O_{65}S_2$: 5200.60; observed: $[M + 4H]^{4+}$ *m/z* = 1301.57, $[M + 5H]^{5+}$ *m/z* = 1041.64, $[M + 6H]^{6+}$ *m/z* = 868.11.

[Cys(Trt-K$_{10}$)^{45}]-TIGIT$^{22–66}$ (13). Peptides **1** (26.6 mg) and **10** (17.2 mg) were dissolved in 1,2,4-triazole ligation buffer (4.2 mL), and the reaction mixture was incubated for 1 h at 37 °C. Then, Trt(OH)-K_{10} (59.7 mg) and TFA (4.2 mL) were added, and the reaction mixture was incubated for 1 h at room temperature. The mixture was purified by preparative HPLC (Cosmosil 5C4-AR300 column, a linear gradient of 18–48% CH_3CN in H_2O containing 0.1% (v/v) TFA over 90 min) to provide **13** (13.4 mg, 35% yield). MS(ESI) *m/z*, calcd for $C_{318}H_{529}N_{97}O_{90}S_3$: 7242.90; observed: $[M + 4H]^{4+}$ *m/z* = 1811.86, $[M + 5H]^{5+}$ *m/z* = 1449.84, $[M + 6H]^{6+}$ *m/z* = 1208.73.

[Cys(Trt-K$_{10}$)^{45}]-TIGIT$^{22–66}$ MPAA ester (14). A $NaNO_2$ solution (0.24 mL) was added to a solution of **13** (22.2 mg) in activation buffer (2.4 mL) at − 20 °C, and the reaction was continued for 30 min. An MPAA solution (0.80 mL) was then added to the reaction, and the reaction mixture was stirred for 5 min at − 20 °C. The mixture was purified by preparative HPLC (Cosmosil Protein-R column, a linear gradient of 25–55% CH_3CN in H_2O containing 0.1% (v/v) TFA over 90 min) to provide **14** (10.0 mg, 52% yield). MS(ESI) *m/z*, calcd for $C_{295}H_{480}N_{78}O_{87}S_4$: 6635.44; observed: $[M + 4H]^{4+}$ *m/z* = 1660.13, $[M + 5H]^{5+}$ *m/z* = 1328.65, $[M + 6H]^{6+}$ *m/z* = 1107.27.

[Thz^{67}]-TIGIT$^{67–141}$ (15). Peptides **11** (20.4 mg) and **4** (20.1 mg) were dissolved in MPAA ligation buffer [100 mM MPAA, 50 mM TCEP, 6 M guanidine, 200 mM phosphate (pH 6.8); 4.0 mL], and the reaction mixture was incubated for 1 h at 37 °C. The mixture was purified by preparative HPLC (Cosmosil 5C18-AR300 column, a linear gradient of 32–62% CH_3CN in H_2O containing 0.1% (v/v) TFA over 90 min) to provide **15** (11.4 mg, 32% yield). MS(ESI) *m/z*, calcd for $C_{413}H_{607}N_{117}O_{118}S_3$: 9189.43; observed: $[M + 7H]^{7+}$ *m/z* = 1314.66, $[M + 8H]^{8+}$ *m/z* = 1150.59, $[M + 9H]^{9+}$ *m/z* = 1022.89, $[M + 10H]^{10+}$ *m/z* = 920.61.

[Thz^{67}, Cys(Trt-K$_{10}$)^{69,107}]-TIGIT$^{67–141}$ (16). Peptide **15** (13.9 mg) was dissolved in reaction buffer [6 M guanidine, 200 mM phosphate (pH 7.0); 1.3 mL], and Trt(OH)-K_{10} (59.7 mg) and TFA (1.3 mL) were added. Then, the reaction mixture was incubated for 1 h at room temperature. The mixture was purified by preparative HPLC (Cosmosil 5C18-AR300 column, a linear gradient of 29–59% CH_3CN in H_2O containing 0.1% (v/v) TFA over 90 min) to provide **16** (9.2 mg, 43% yield). MS(ESI) *m/z*, calcd for $C_{581}H_{887}N_{161}O_{146}S_3$: 12551.61; observed: $[M + 12H]^{12+}$ *m/z* = 1047.80, $[M + 13H]^{13+}$ *m/z* = 967.33, $[M + 14H]^{14+}$ *m/z* = 898.27, $[M + 15H]^{15+}$ *m/z* = 838.52.

[Cys^{67}, Cys(Trt-K$_{10}$)^{69,107}]-TIGIT$^{67–141}$ (17). Peptide **16** (5.7 mg) was dissolved in reaction buffer [6 M guanidine, 200 mM phosphate (pH 7.0); 0.35 mL], and 2 M $MeONH_2$ (39 μL) was added. Then, the reaction mixture was incubated for 2 h at 37 °C. The mixture was purified by preparative HPLC (Cosmosil 5C18-AR300 column, a linear gradient of 31–61% CH_3CN in H_2O containing 0.1% (v/v) TFA over 90 min) to provide **17** (4.7 mg, 81% yield). MS(ESI) *m/z*, calcd for $C_{580}H_{887}N_{161}O_{146}S_3$: 12539.61; observed: $[M + 12H]^{12+}$ *m/z* = 1046.71, $[M + 13H]^{13+}$ *m/z* = 966.22, $[M + 14H]^{14+}$ *m/z* = 897.35, $[M + 15H]^{15+}$ *m/z* = 837.61.

[Cys(Trt-K$_{10}$)^{45,69,107}, Cys^{67}]-TIGIT$^{22–141}$ (18). Peptides **14** (6.2 mg) and **17** (6.3 mg) were dissolved in MPAA ligation buffer (0.76 mL), and the reaction mixture was incubated for 1 h at 37 °C. The mixture was purified by preparative HPLC (Cosmosil 5C4-AR300 column, a linear gradient of 25–55% CH_3CN in H_2O containing 0.1% (v/v) TFA over 90 min) to provide **18** (7.4 mg, 79% yield). MS(ESI) *m/z*, calcd for $C_{867}H_{1359}N_{239}O_{231}S_6$: 19007.03; observed: $[M + 14H]^{14+}$ *m/z* = 1359.04, $[M + 15H]^{15+}$ *m/z* = 1268.40, $[M + 16H]^{16+}$ *m/z* = 1189.27, $[M + 17H]^{17+}$ *m/z* = 1119.36.

[Cys(Trt-K$_{10}$)^{45,69,107}]-TIGIT$^{22–141}$ (19). Peptides **18** (7.4 mg) was dissolved in desulfurization buffer [20 mM VA-044, 100 mM MESNa, 250 mM TCEP, 6 M guanidine, 100 mM phosphate (pH 6.5); 1.5 mL], and the reaction mixture was incubated for 2 h at 37 °C. The mixture was purified by preparative HPLC (Cosmosil 5C4-AR300 column, a linear gradient of 27–57% CH_3CN in H_2O containing 0.1% (v/v) TFA over 90 min) to provide **19** (2.7 mg, 38% yield). MS(ESI) *m/z*, calcd for $C_{867}H_{1359}N_{239}O_{231}S_5$: 18975.05; observed: $[M + 14H]^{14+}$ *m/z* = 1356.82, $[M + 15H]^{15+}$ *m/z* = 1266.20, $[M + 16H]^{16+}$ *m/z* = 1187.34, $[M + 17H]^{17+}$ *m/z* = 1117.46.

TIGIT$^{22–141}$ (20). Peptides **19** (0.57 mg) was dissolved in TFA/TIS (95:5) and the solution was incubated for 30 min at room temperature. The reaction mixture was poured into ice-cold Et_2O. the precipitate was dissolved in denaturation buffer [6 M guanidine, 20 mM Tris (pH 8.0); 0.24 mL] and purified by preparative SEC [TSKgel G2000SW$_{XL}$, 1 mM DTT, 6 M guanidine, 20 mM Tris (pH 8.0); 0.5 mL/min] to provide **20** (20.6 μM, 0.40 mL, 41%). Peptide **20** was quantified by UV absorbance at 280 nm, and the purity of **20** was confirmed by SDS-PAGE analysis. MS(ESI) *m/z*, calcd for $C_{615}H_{939}N_{173}O_{189}S_5$: 13931.78; observed: $[M + 11H]^{11+}$ *m/z* = 1267.53, $[M + 12H]^{12+}$ *m/z* = 1162.82, $[M + 14H]^{14+}$ *m/z* = 996.44.

Folding of TIGIT$^{22–141}$. Purified TIGIT$^{22–141}$ (~ 20 μM) in guanidine solution [1 mM DTT, 6 M guanidine, 20 mM Tris (pH 8.0); 0.14 mL] was diluted into folding buffer [400 mM arginine, 3 mM EDTA, 5 mM reduced glutathione, 0.5 mM oxidized glutathione, 150 mM NaCl, 10 mM HEPES (pH 7.4); 10 mL] and incubated for 18 h at 4 °C. The protein solution was concentrated and buffer exchange to HEPES buffer [3 mM EDTA, 150 mM NaCl, 10 mM HEPES (pH 7.4)] was performed by ultrafiltration using an Amicon ultra centrifugation filter (NMWL 3 kDa, Millipore). The protein solution was filtered and applied to the following biological assay.

CD spectral analysis. Refolded TIGIT$^{22–141}$ (~ 6 μM) in PBS was used for measurement. CD spectrum was recorded on a J-720 circular dichroism spectrometer (JASCO) at 20 °C.

SPR analysis. SPR analysis of synthetic TIGIT$^{22–141}$ to CD155 was carried out using a Biacore X-100 (Cytiva). After immobilizing biotinylated human CD155/PVR (ACROBiosystems) on a sensor chip SA, the analyte protein solution of synthetic TIGIT$^{22–141}$ was flashed into the flow system in the running buffer [10 mM HEPES (pH 7.4), 150 mM NaCl, 3 mM EDTA, 0.005% Surfactant P20] at 25 °C. All analytes were evaluated for 90 s as contact time, followed by 600 s dissociation at a 30 μL/min flow rate. One molar NaCl was used for regeneration. The triplicate data were analyzed using GraphPad Prism software.

References

1. Yu X et al (2009) The surface protein TIGIT suppresses T cell activation by promoting the generation of mature immunoregulatory dendritic cells. Nat Immunol 10:48–57
2. Chauvin JM, Zarour HM (2020) TIGIT in cancer immunotherapy. J Immunother Cancer 8:e00957
3. Florou V, Garrido-Laguna I (2022) Clinical development of anti-TIGIT antibodies for immunotherapy of cancer. Curr Oncol 24:1107–1112
4. Stengel KF et al (2012) Structure of TIGIT immunoreceptor bound to poliovirus receptor reveals a cell-cell adhesion and signaling mechanism that requires cis-trans receptor clustering. Proc Natl Acad Sci USA 109:5399–5404
5. Manieri NA et al (2017) TIGIT: A key inhibitor of the cancer immunity cycle. Trends Immunol 38:20–28
6. Jeong BS et al (2022) Structural and functional characterization of a monoclonal antibody blocking TIGIT. mAbs 14:e2013750
7. Diong SJ et al (2023) Biophysical characterization of PVR family interaction and therapeutic antibody recognition to TIGIT. mAbs 15:2253788
8. Zhou X et al (2020) A novel D-peptide identified by mirror-image phage display blocks TIGIT/PVR for cancer immunotherapy. Angew Chem Int Ed 59:15114–15118
9. Zheng JS et al (2016) Robust chemical synthesis of membrane proteins through a general method of removable backbone modification. J Am Chem Soc 138:3553–3561
10. Zhang B et al (2023) Backbone-installed split intein-assisted ligation for the chemical synthesis of mirror-image proteins. Angew Chem Int Ed 62:e202306270
11. Tsuda S et al (2018) Easy-to-attach/detach solubilizing-tag-aided chemical synthesis of an aggregative capsid protein. Angew Chem Int Ed 57:2105–2109
12. Dawson PE et al (1994) Synthesis of proteins by native chemical ligation. Science 266:776–779

13. Blanco-Canosa JB et al (2015) Chemical protein synthesis using a second-generation *N*-acylurea linker for the preparation of peptide-thioester precursors. J Am Chem Soc 137:7097–7209
14. Dang B et al (2013) Native chemical ligation at Asx-Cys, Glx-Cys: Chemical synthesis and high-resolution X-ray structure of ShK toxin by racemic protein crystallography. J Am Chem Soc 135:11911–11919
15. Blanco-Canosa JB, Dawson PE (2008) An efficient Fmoc-SPPS approach for the generation of thioester peptide precursors for use in native chemical ligation. Angew Chem Int Ed 47:6851–6855
16. Sakamoto K et al (2017) 1,2,4-Triazole-aided native chemical ligation between peptide-*N*-acyl-*N*'-methyl-benzimidazolinone and cysteinyl peptide. Chem Commun 53:12236–12239
17. Wang JX et al (2015) Free-radical-based, specific desulfurization of cysteines: A powerful advance in the synthesis of polypeptides and glycopolypeptides. Angew Chem Int Ed 54:2194–2198
18. Hackeng TM et al (1999) Protein synthesis by native chemical ligation: Expanded scope by using straightforward methodology. Proc Natl Acad Sci USA 96:10068–10073
19. Tsuda S et al (2019) The versatile use of solubilizing trityl tags for difficult peptide/protein synthesis. Org Biomol Chem 17:1202–1205
20. Wan Q, Danishefsky SJ (2007) Free-radical-based, specific desulfurization of cysteines: A powerful advance in the synthesis of polypeptides and glycopolypeptides. Angew Chem Int Ed 46:9248–9252
21. Samanta D et al (2017) Structural, mutational and biophysical studies reveal a canonical mode of molecular recognition between immune receptor TIGIT and nectin-2. Mol Immunol 81:151–159
22. Deuss FA et al (2017) Recognition of nectin-2 by the natural killer cell receptor T cell immunoglobulin and ITIM domain (TIGIT). J Biol Chem 292:11413–11422

Chapter 3
Mirror-Image Human Serum Albumin Domain III as a Tool for Analyzing Site II-Dependent Molecular Recognition

Abstract Human serum albumin (HSA) as a drug carrier can significantly improve the pharmacokinetic profiles of short-lived therapeutics. Conjugation of albumin-binding moieties (ABMs) to therapeutic agents may prolong their serum half-life by promoting their association with endogenous HSA. To discover a new molecular class of ABMs from mirror-image chemical space, a preparation protocol for bioactive HSA domain III and its D enantiomer (D-HSA domain III) was established. Structural and functional analyses suggested that the synthetic protein enantiomers exhibited mirror-image structures and stereoselective neonatal fragement crystallizable receptor (FcRn) recognition. Additionally, the ligand-binding properties of synthetic L-HSA domain III were comparable with those of site II in native HSA, as confirmed using site II-selective fluorescent probes and an esterase substrate. Synthetic D-HSA domain III is an attractive tool for analyzing the site II-dependent molecular recognition properties of HSA.

Keywords Human serum albumin domain III · Albumin binding moiety · Chemical protein synthesis · Mirror-image protein · Site II · Neonatal fragement crystallizable receptor · Dansyl amino acid

Human serum albumin (HSA) is the most abundant protein in human blood, with a concentration of 35–50 g/L, and is critically involved in regulation of plasma pH and oncotic pressure in human physiology [1]. HSA binds to a wide variety of molecular ligands and thus serves as an important carrier of endogenous and exogenous substances [2]. Importantly, HSA has an extraordinarily long serum half-life of 19–21 d owing to its unique physicochemical properties and intrinsic molecular interactions [3]. The relatively large molecular mass (66.5 kDa) and net negative charge of HSA at physiological pH hamper its elimination from the circulation by glomerular filtration. In addition, its pH-dependent interaction with the neonatal fragment crystallizable receptor (FcRn) allows HSA to escape lysosomal degradation [3].

HSA consists of three globular domains (denoted I, II and III) that collectively form an asymmetric heart-shape structure (Fig. 3.1a) [4]. Although these domains

N. Iwamoto, *Synthesis of Functional Characterization of Protein Domains for the Development of Mirror-Image Therapeutics*, Springer Theses,
https://doi.org/10.1007/978-981-95-5366-2_3

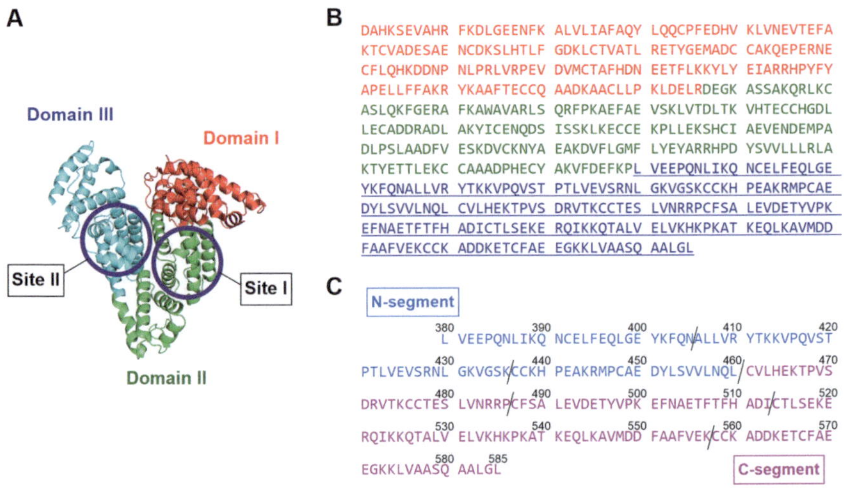

Fig. 3.1 Human serum albumin (HSA) and its three domains. **a** Structure of HSA, with domain shown in different colors. Drug binding sites I and II are contained within the encircled regions. **b** Sequence of HSA (585 amino acids). The sequence of domain III is underlined. **c** Design of seven peptide segments for chemical synthesis of HSA domain III (HSA$^{380-585}$). Two synthetic intermediates, N-segment and C-segment, are shown in cyan and magenta, respectively

share homologous α-helix-rich regions, each domain has characteristic biological functions, such as small-molecule binding and FcRn binding [5, 6]. Two representative drug binding sites in HSA, namely site I (warfarin site) and site II (diazepam site), are mainly attributed to interactions with domains II and III, respectively. Domain III is also responsible for the interaction with FcRn, together with the partial involvement of domain I. Structural and functional analyses of individual domains have provided deep insights into their physiological roles and relevant molecular interactions [7–9].

Practical medicinal applications of HSA include its use as a drug carrier [10]. Taking advantage of its high solubility and long serum half-life, covalent or noncovalent association with HSA greatly improves the physicochemical and pharmacokinetic profiles of cargo molecules. For example, administration of nanoparticle albumin-bound paclitaxel results in a greater fraction of unbound paclitaxel compared with conventional solvent-based delivery methods, together with efficient transport of albumin-bound paclitaxel into tumor tissues [11]. Additionally, HSA conjugation of glucagon-like-peptide 1 significantly extends its serum half-life [12]. Alternatively, therapeutic agents can be genetically or chemically appended to albumin-binding moieties (ABMs) that bind to endogenous HSA upon in vivo administration [13]. Ozoralizumab is a bispecific treatment for rheumatoid arthritis, consisting of two human-TNFα-binding variable domains of heavy chains of heavy-chain antibodies (VHHs) and one HSA-binding VHH [14]. Binding to HSA provides ozoralizumab with a long serum half-life of 18 d, which is comparable to that of a conventional immunoglobulin G antibody. The distinct localization of ozoralizumab to inflamed joints in arthritic model mice was also reported, presumably because albumin preferentially accumulates at sites of inflammation [15]. Thus, for the purpose of improving the pharmacokinetics of bioactive ingredients, ABMs with various molecular structures have been developed, including small molecules, peptides, and antibody domains [16–18]. The favorable pharmacokinetic profiles of ABMs, which are attributable to their molecular size, epitope, thermal stability, and HSA binding affinity, will thus also apply to their appended therapeutics [13, 19].

As described in Introduction, the symmetrical properties of D-proteins provide intriguing tools to explore potential D-peptide- and D-protein-based therapeutics [20], which represent promising protease-resistant drug scaffolds with low immunogenicity [21–23]. The author speculated that albumin-binding D-peptides and D-proteins may provide a promising molecular class of ABMs. D-Peptide- and D-protein-based ABMs should have higher in vivo stability and lower immunogenicity than their corresponding L enantiomers while retaining the favorable characteristics of the latter, such as specificity, affinity and thermal stability. To explore D-peptide/protein-based ABMs, preparation of the bioactive D enantiomer of HSA (D-HSA) is needed [24]. In this study, the author therefore established a protocol for the synthesis of D-HSA domain III. The author then analyzed the site II-dependent ligand interactions of the synthetic L- and D-HSA domain III proteins.

Synthesis of HSA Domain III. $HSA^{380–585}$ (domain III) was chosen as a synthetic target in this study (Fig. 3.1b). Functional domain III protein can be efficiently obtained by refolding recombinantly expressed protein [8]. Additionally, the isolated domain retains several important biological functions, such as FcRn binding and site II-dependent ligand interactions [7, 9], which enable functional evaluations of the synthetic protein. The author designed a strategy for synthesizing $HSA^{380–585}$ from two large peptide segments, N-segment ($HSA^{380–460}$) and C-segment ($HSA^{461–585}$), which can be subsequently ligated to form a Leu460–Cys461 peptide bond using native chemical ligation (NCL) (Fig. 3.1c) [25]. Because the Leu–Cys junction has minimal steric hindrance, the author expected the final ligation between the N- and C-segments to proceed efficiently [26].

For the synthesis of N-segment L-**8**, three peptide subsegments (L-**1**–L-**3**) were designed and synthesized (Fig. 3.2). All peptide subsegments were constructed using Fmoc-based solid-phase peptide synthesis (Fmoc-SPPS) on a hydrazide-linked resin [27, 28]. The resulting peptide hydrazide was directly subjected to acetylacetone (acac)-mediated hydrazide activation [29] to yield the N-terminal peptide segment ($HSA^{380–405}$) as a 4-mercaptophenylacetic acid (MPAA) thioester L-**1**. The side-chain of Cys392 in L-**1** was protected with an acetamidomethyl (Acm) group. In a previous report, preparation of a peptide hydrazide with C-terminal Asn on a hydrazide carboxylate Wang resin failed due to the intramolecular cyclization between the hydrazide and the Asn side-chain amide during SPPS [27]. In this study, use of Trt(2-Cl) resin [28] prevented this unfavorable reaction, leading to the successful preparation of the peptide hydrazide with C-terminal Asn. Because the spontaneous formation of the cyclization product was observed during the peptide cleavage and storage, the resulting peptide hydrazide was immediately used to the subsequent thioester formation for L-**1**. Peptides L-**1** and L-**2** ($HSA^{406–436}$) were subjected

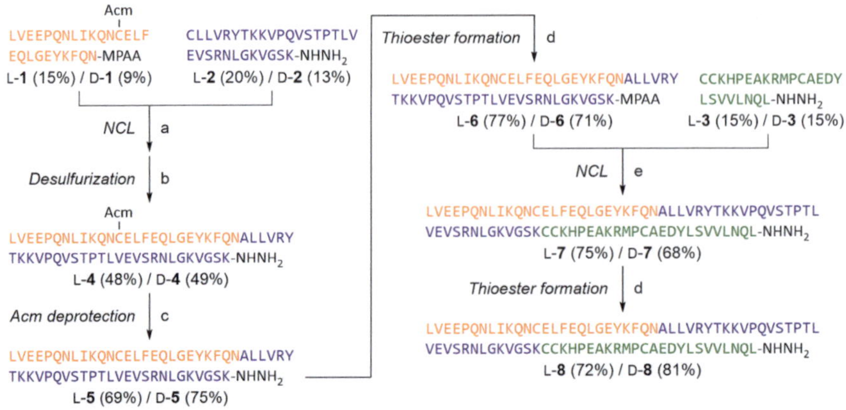

Fig. 3.2 Synthesis of the N-segment (HSA$^{380–460}$). *Reagents and conditions*: **a** 2.5 M 1,2,4-triazole, 30 mM TCEP, 6 M guanidine, 100 mM phosphate (pH 7); **b** 25 mM VA-044, 40 mM MESNa, 240 mM TCEP, 6 M guanidine, 100 mM phosphate (pH 6.5); **c** AgOTf, anisole, TFA; then 10 mM DTT, 6 M guanidine PBS; **d** acac, 200 mM MPAA, 6 M guanidine; **e** 100 mM MPAA, 50 mM TCEP, 6 M guanidine, 200 mM phosphate (pH 7)

to 1,2,4-triazole-mediated NCL [30] followed by one-pot desulfurization [31] to provide the ligated peptide L-**4**. Silver trifluoromethanesulfonate-mediated deprotection of the Acm group [32] in L-**4** provided peptide hydrazide L-**5**, which was further converted into thioester L-**6**. Peptide L-**6** and peptide L-**3** (HSA$^{437–460}$) were subjected to MPAA-mediated NCL to yield L-**7**, a peptide hydrazide of the desired N-segment (HSA$^{380–460}$). To minimize the unfavorable lactam formation from thioester L-**6** with the C-terminal Lys [33], the excess amount of L-**3** was used for the rapid completion of the NCL (~ 1 h), providing the NCL product L-**7** in satisfactory yield (75% from peptide L-**6**). The C-terminal hydrazide in L-**7** was converted into an MPAA thioester in L-**8** (N-segment).

Next, the author investigated the synthesis of C-segment L-**16** from four peptide subsegments (L-**9**–L-**12**) (Fig. 3.3). MPAA thioester of peptide L-**9** (HSA$^{461–486}$) was constructed by acac-mediated activation of the corresponding peptide hydrazide. Peptide L-**10** (HSA$^{487–513}$) was constructed on a hydrazide-linked resin using Fmoc-SPPS. Peptide L-**11** (HSA$^{514–557}$) was constructed on an *N*-acyl-*N'*-methylbenzimidazolinone (MeNbz)-linked resin [34] with a C-terminal poly-arginine tag, which improved the solubility and thus facilitated the purification of this segment. The N-terminal Cys461 and Cys514 residues in peptides L-**9** and L-**11**, respectively, were each protected as a 1,3-thiazolidine-4-carboxylic acid (Thz). Peptide L-**12** (HSA$^{558–585}$) was synthesized with a C-terminal (+)-biotin tag for immobilization during the functional analysis and library screening. Initially, NCL of peptides L-**9** and L-**10** was performed, but limited formation of a Pro486–Cys487 junction occurred under the standard MPAA-mediated NCL conditions [26]. After several investigations, the author found that a ligation buffer with a relatively high tris(2-carboxyethyl)phosphine (TCEP) concentration [100 mM MPAA, 100 mM TCEP,

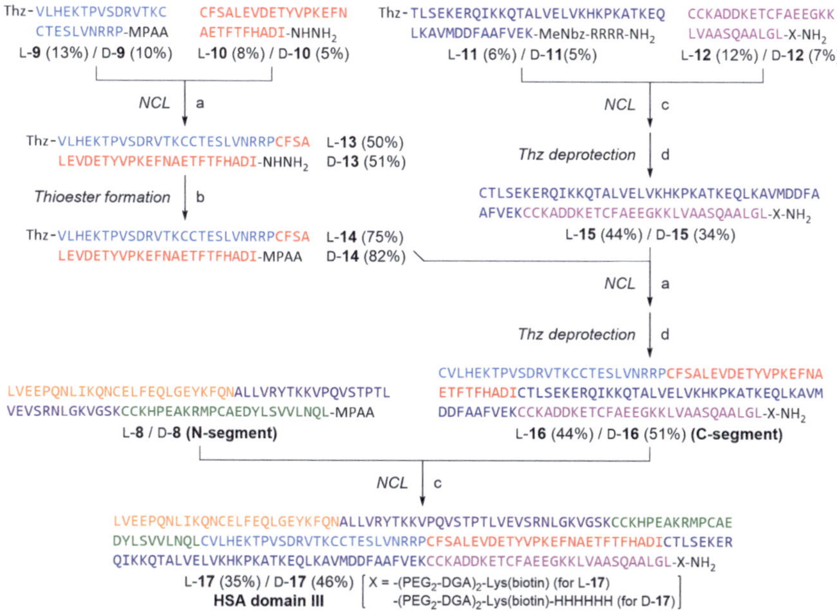

Fig. 3.3 Synthesis of HSA domain III (HSA[380–585]). *Reagents and conditions*: **a** 100 mM MPAA, 100 mM TCEP, 6 M guanidine, 200 mM phosphate (pH 7); **b** acac, 200 mM MPAA, 6 M guanidine; **c** 100 mM MPAA, 50 mM TCEP, 6 M guanidine, 200 mM phosphate (pH 7); **d** methoxyamine

6 M guanidine, 200 mM phosphate (pH 7)] and an extended reaction time (18 h) at 37 °C provided a satisfactory yield (50%) of the ligated product L-**13**. During the reaction, formation of deletion products [35] was not observed, presumably because the C-terminal Arg–Pro thioester was less prone to diketopiperazine formation [36]. The acac-mediated thioester formation from hydrazide L-**13** provided the MPAA thioester L-**14** without modification of the N-terminal Thz moiety. Separately, peptides L-**11** and L-**12** were subjected to MPAA-mediated NCL followed by one-pot methoxyamine-mediated deprotection of the Thz group to provide peptide L-**15** (44%, 2 steps from peptides L-**12**). NCL of L-**14** and L-**15** to form an Ile513–Cys514 peptide bond was also inefficient [26]. To optimize the reaction conditions, thioester L-**14** was added to the reaction mixture in two portions, and an extended reaction time (15 h) in MPAA ligation buffer with increased TCEP concentration was also needed. Because peptide L-**15** could not be separated from the ligated product using preparative HPLC, complete consumption of L-**15** during NCL was critical. After one-pot deprotection of the Thz group at Cys460, C-segment L-**16** (HSA[461–585]) was isolated with a sufficient yield (44%).

The author finally investigated the assembly of N-segment L-**8** and C-segment L-**16** (Fig. 3.3). NCL of L-**8** and L-**16** in MPAA ligation buffer proceeded smoothly and provided a 35% yield of full-length HSA domain III (HSA[380–585], L-**17**). Based on the synthetic protocols established, the D enantiomer of HSA[380–585] (D-**17**) with a C-terminal poly-L-histidine tag and a (+)-biotin tag was successfully synthesized from N-segment D-**8** and C-segment D-**16** in a similar manner. The overall synthetic yields of $HSA^{380–585}$ (L-**17**: 6.8% from L-**12**; D-**17**: 8.0% from D-**12**) were satisfactory to provide the materials for the subsequent experiments.

Structural and Biological Evaluation of Synthetic HSA Domain III. The author next investigated the refolding of L-$HSA^{380–585}$ and D-$HSA^{380–585}$ to obtain bioactive HSA domain III enantiomers. Domain III contains twelve cysteine residues, all of which are involved in disulfide bond formation [1]. As reported previously [8, 37], refolding proceeded under weakly basic conditions. The denatured proteins were diluted and dialyzed against carbonate buffer (pH 10), and then the partially-refolded proteins were oxidatively refolded using cystamine. The formation of disulfide bonds was confirmed using mass spectrometry after iodoacetamide-mediated alkylation of free thiol groups (Fig. 3.4). For successful refolding, the use of cystamine as an oxidant was critical as previously reported [7, 38]. The circular dichroism (CD) spectrum of refolded L-HSA domain III showed features consistent with an α-helix-rich structure and was similar to that of recombinant HSA domain III (Fig. 3.5a). Additionally, the CD spectrum of refolded D-HSA domain III was inverted relative to that of the L-isomer, consistent with their mirror-image structures.

The biological functions of the synthetic proteins were investigated as described previously [9]. The binding of HSA domain III to FcRn was evaluated using surface plasmon resonance (SPR) under slightly acidic conditions (pH 5.5) mimicking the endosomal environment [6]. The synthetic L-HSA domain III bound to FcRn with a micromolar dissociation constant ($K_D = 8.0 \pm 0.9\ \mu M$), which was comparable to the one in a previous study using recombinantly-expressed HSA domain III ($K_D = 3.2\ \mu M$) [9]. Meanwhile, no binding of D-HSA domain III was observed (Fig. 3.5b), which suggested the stereoselective interaction between the HSA enantiomers and native FcRn.

Molecular Recognition of Synthetic L- and D-HSA Domain III Proteins. HSA includes two distinct drug binding sites. Site I is located in domain II, which recognizes negatively-charged heterocyclic molecules, such as warfarin. Site II in domain III preferentially binds to aromatic carboxylic acids, such as ibuprofen, which contain negatively-charged acidic moieties in addition to hydrophobic aromatic rings [39]. In previous studies, HSA–drug interactions at site I and II have been assessed using various methodologies, including crystallography [40], nuclear magnetic resonance spectroscopy [41], CD spectroscopy [42], and fluorescence quenching [43]. Fluorescent probe-based assays, in which HSA–drug interactions are monitored using HSA-specific fluorescent probes, also provide a simple analytical approach [44]. The author speculated that these highly sensitive fluorescence assays would be suitable for evaluating the interactions between synthetic HSA domain III and known HSA-binding molecules.

Several fluorescent probes with different site selectivities and emission wavelengths have been developed for the detection of HSA binding [45, 46]. These probes have a high quantum yield upon HSA binding, facilitating the quantification of HSA

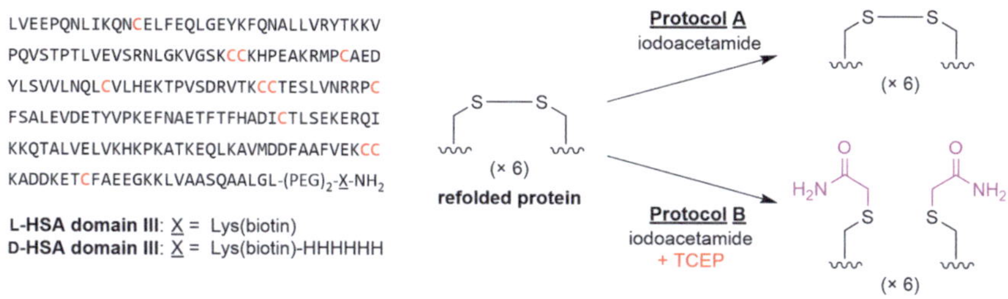

Fig. 3.4 Confirmation of disulfide bond formation in refolded L- and D-HSA domain III proteins. After incubation of each protein with iodoacetamide in the absence (protocol A) or presence (protocol B) of TCEP, the product was analyzed using LC–MS. Calculated masses and observed *m/z* values for each product are shown in the table

Protein	Protocol	Formula	Calcd	Observed *m/z*
L-HSA domain III	A	$C_{1066}H_{1724}N_{285}O_{324}S_{15}$	24181.29	$[M + 16H]^{16+}$ = 1512.85 $[M + 17H]^{17+}$ = 1423.70
	B	$C_{1090}H_{1772}N_{297}O_{336}S_{15}$	24877.64	$[M + 16H]^{16+}$ = 1556.35 $[M + 17H]^{17+}$ = 1464.80
D-HSA domain III	A	$C_{1102}H_{1766}N_{303}O_{330}S_{15}$	25003.65	$[M + 16H]^{16+}$ = 1564.35 $[M + 17H]^{17+}$ = 1472.50
	B	$C_{1126}H_{1814}N_{315}O_{342}S_{15}$	25700.00	$[M + 16H]^{16+}$ = 1608.00 $[M + 17H]^{17+}$ = 1513.50

in biological samples and the characterization of site-selective interactions of therapeutic agents. For functional evaluation of synthetic HSA domain III, the author used dansyl amino acids [47, 48]. The HSA-binding properties of dansyl amino acids depend on their side-chain structures and stereochemistries [49, 50]. Initially, achiral dansyl-Gly, a site II-selective probe, was used to confirm the existence of a functional drug binding pocket in the synthetic proteins. Addition of increasing concentrations of L-HSA domain III (0–4 μM) to dansyl-Gly (10 μM; excitation at 365 nm, emission at 480 nm) led to a corresponding increase in the fluorescence emission intensity (Fig. 3.6a). As expected, similar results were obtained using D-HSA domain III and recombinant HSA domain III. In contrast, no fluorescence signal was detected when a negative control, human ubiquitin, was added to dansyl-Gly. These results suggested that both synthetic HSA domain III enantiomers reproduced the specific functions of recombinant HSA domain III.

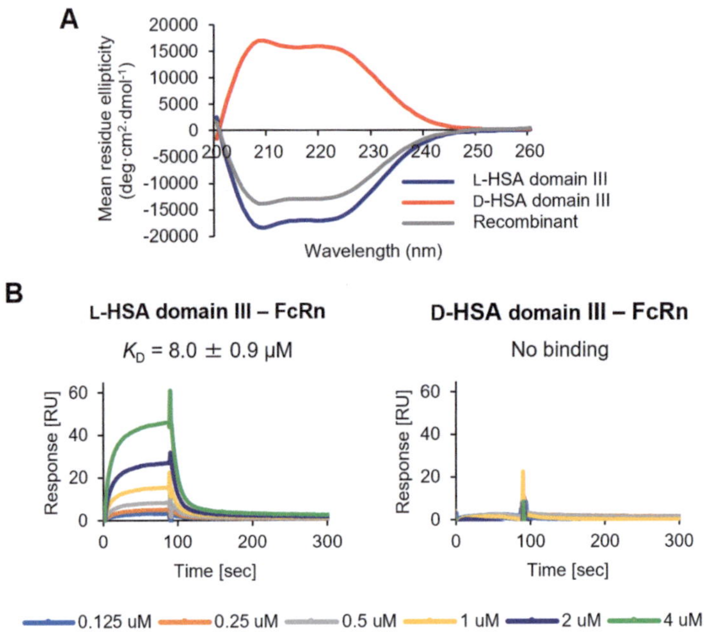

Fig. 3.5 Characterization of L- and D-HSA domain III proteins. **a** CD spectra of refolded synthetic L- and D-HSA domain III and recombinant HSA domain III. **b** SPR sensorgrams of the association, steady-state binding, and dissociation interactions of synthetic HSA domain III proteins and fragment crystallization receptor (FcRn). Various concentrations of L- or D-HSA domain III were injected over immobilized human FcRn (≈ 1100 RU). The binding affinity was determined from triplicate assays by fitting a 1:1 binding model

Dansyl amino acids with hydrophobic side-chains were reported to selectively bind to HSA site II [49]. To investigate the stereoselectivity of the ligand–HSA site II interactions, the author evaluated the binding of several chiral dansyl amino acids to L- and D-HSA domain III proteins by assessing the intensity of dansyl fluorescence emission at 480 nm. The fluorescence intensity of each dansyl amino acid (3 μM) in the presence of HSA domain III (3 μM) was then calculated relative to that of dansyl-Gly. In the presence of L-HSA domain III, the fluorescence emission intensity of dansyl-L-Phe was three times that of dansyl-Gly, while the emission intensity of dansyl-D-Phe was twice that of dansyl-Gly. Similarly, in the presence of L-HSA domain III, dansyl-L-Pro and dansyl-L-Val had higher fluorescence emission intensities than with dansyl-D-Pro and dansyl-D-Val, respectively (Fig. 3.6b). Comparable binding profiles were also obtained for recombinant HSA domain III (Fig. 3.6b). These findings suggested that L-HSA domain III preferentially binds to the L-enantiomers of dansyl amino acids. In a previous report on HPLC analysis, site II-binding dansyl L-amino acids were eluted from an HSA-immobilized chiral column with a shorter retention time, which might suggest the lower affinity of dansyl L-amino acids with full-length HSA [50]. The discrepancy may be due to differences

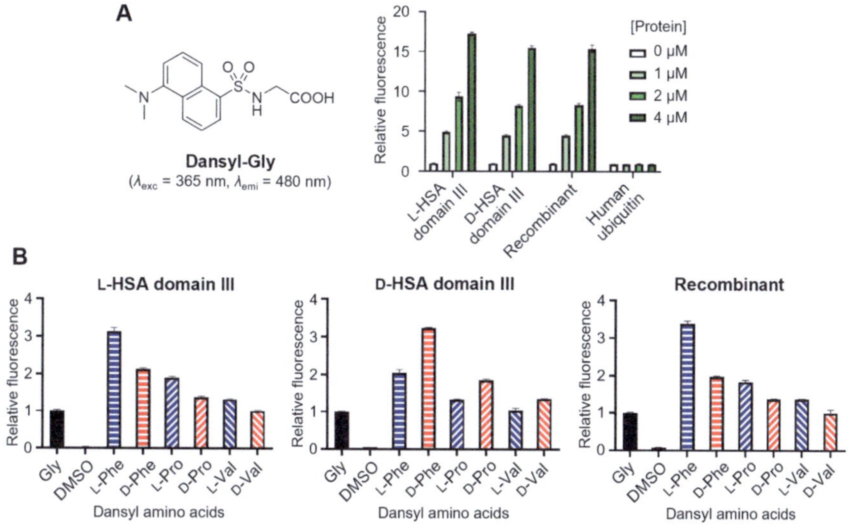

Fig. 3.6 Ligand-binding properties of synthetic HSA domain III. **a** Structure of dansyl-Gly, and its fluorescence excitation (λ_{exc}) and emission (λ_{emi}) wavelengths upon binding to L- or D-HSA domain III. Dansyl-Gly (10 μM) was incubated with synthetic HSA domain III or recombinant HSA domain III (1–4 μM) in phosphate-buffered saline (PBS) for 30 min. Human ubiquitin was used as a negative control. The fluorescence emissions were normalized relative to that measured in the absence of the proteins. Mean ± standard deviation ($n = 3$) are shown. **b** Fluorescence emission intensity of chiral dansyl amino acids after binding to HSA domain III. Dansyl amino acids (3 μM) were incubated with synthetic HSA domain III or recombinant HSA domain III (3 μM) in PBS for 30 min. The emission intensity for each dansyl amino acid was normalized relative to that for dansyl-Gly

in analytical methods or the molar proportions of added dansyl amino acids. For D-HSA site II, the binding profiles of chiral dansyl amino acids were the opposite of those observed for L-HSA; namely, D-Phe, D-Pro, and D-Val exhibited higher fluorescence intensities in the presence of D-HSA domain III than L-Phe, L-Pro, and L-Val, respectively (Fig. 3.6b).

To further demonstrate the validity and applicability of the synthetic HSA domain III proteins, a competition assay to identify potential site II ligands was designed. (±)-Ibuprofen, a site II-binding drug, inhibited the interaction between dansyl-Gly and L-HSA domain III in a dose-dependent manner, as evidenced by the dose-dependent decrease in dansyl-Gly fluorescence (Fig. 3.7a). In contrast, (±)-warfarin, a site-I binding drug, did not inhibit this interaction. Similar results were obtained using D-HSA domain III. Of note, no stereoselectivity of the inhibition by (*R*)- and (*S*)-ibuprofen was observed, indicating that both ibuprofen enantiomers efficiently associated with HSA domain III regardless of the chirality at the α-position of the propionate moiety (Fig. 3.7b). Overall, these results indicated that synthetic HSA domain III proteins had the same ligand-binding properties as native HSA site II.

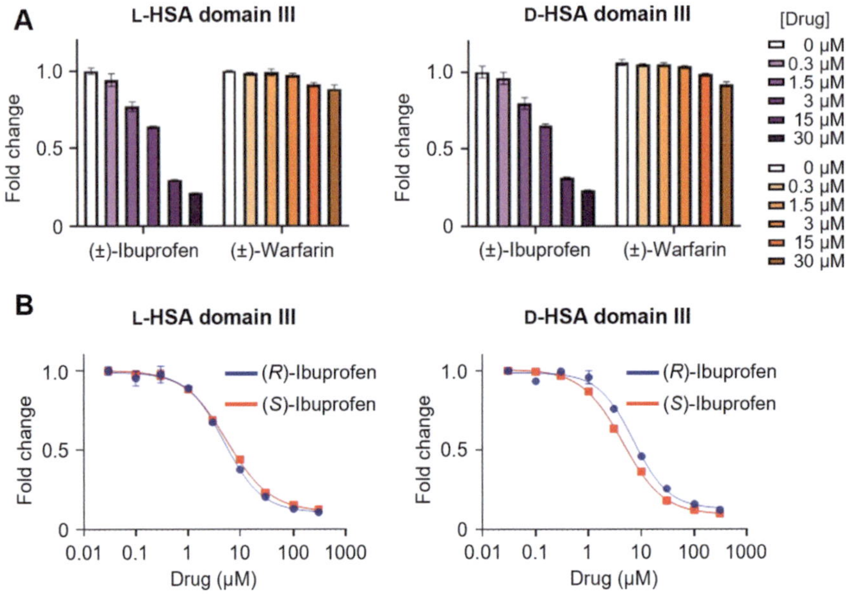

Fig. 3.7 Inhibition of dansyl-Gly binding to HSA domain III. **a** Inhibition of dansyl-Gly binding to HSA domain III by (±)-ibuprofen (purple) or (±)-warfarin (orange). Dansyl-Gly (3 μM) was incubated with L- or D-HSA domain III (3 μM) in the presence of (±)-ibuprofen or (±)-warfarin (0.3–30 μM). The fluorescence emissions of dansyl-Gly in each case was normalized relative to that measured in the absence of competing drugs. **b** Inhibition of dansyl-Gly binding to HSA domain III by (*R*)-ibuprofen (blue) or (*S*)-ibuprofen (red). Dansyl-Gly (3 μM) was incubated with L- or D-HSA domain III (3 μM) in the presence of (*R*)-ibuprofen or (*S*)-ibuprofen (0.03–300 μM) in PBS for 30 min. The fluorescence emissions of dansyl-Gly in each case was normalized relative to that measured in the presence of 0.03 μM of drugs

Esterase-like Activity of Synthetic HSA Domain III. HSA was reported to show esterase-like activity at each drug binding site [51, 52]. *p*-Nitrophenyl acetate (PNPA) is a representative substrate for site II, which is hydrolyzed by turnover at Tyr411 to give *p*-nitrophenol (PNP) and acetic acid (Fig. 3.8a) [52, 53]. The hydrolysis proceeds in two steps: acetylation of the phenolic hydroxy group of Tyr411 with PNPA and hydrolysis of acetylated Tyr411. The initial acetylation proceeds more rapidly than the hydrolysis, accounting for the biphasic kinetic profile of PNP formation during HSA-mediated PNPA hydrolysis [53]. The initial rate of PNP formation is mainly attributed to the rapid acetylation of catalytic Tyr411, even in the presence of a large excess of PNPA [52]. In several reports, the hydrolysis of PNPA by HSA has been assessed using a first-order rate constant [54, 55] to confirm the moderate esterase-like activity of recombinant HSA domain III [55].

To investigate the esterase-like activity of synthetic HSA domain III, the kinetic profiles of PNPA hydrolysis were monitored using the UV absorbance of PNP at 400 nm (Fig. 3.8b, Table 3.1). In the presence of L-HSA domain III (5 μM) in 100 mM phosphate buffer (pH 8), PNPA (100 μM) hydrolysis followed an observed

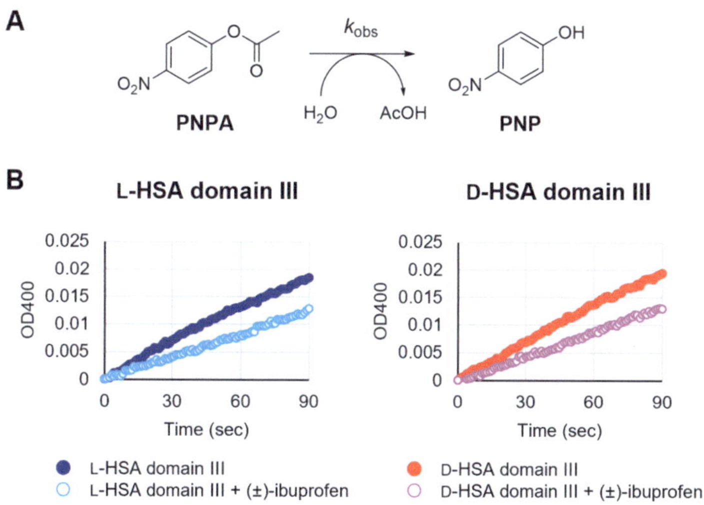

Fig. 3.8 Esterase-like activity of synthetic HSA domain III. **a** HSA-mediated hydrolysis of *p*-nitrophenyl acetate (PNPA), yielding *p*-nitrophenol (PNP) via an observed first-order rate constant k_{obs}. **b** Time course of PNP formation during hydrolysis of PNPA (100 μM) by L- or D-HSA domain III (5 μM) in the presence or absence of (±)-ibuprofen (10 μM) in 100 mM phosphate buffer (pH 7.9) at 25 °C. Product formation was monitored by measuring the UV absorbance of PNP at 400 nm

Table 3.1 Observed first-order kinetic constant (k_{obs}) of *p*-nitrophenyl acetate hydrolysis by human serum albumin (HSA) domain III

Conditions	k_{obs} (10^{-4} s^{-1})[a]
L-HSA domain III	1.11 ± 0.02
L-HSA domain III + (±)-ibuprofen	0.77 ± 0.04
D-HSA domain III	1.18 ± 0.03
D-HSA domain III + (±)-ibuprofen	0.82 ± 0.06

[a] k_{obs} values were determined from triplicate assays

first-order rate constant $k_{obs} = [1.11 \pm 0.02] \times 10^{-4}$ s^{-1}. D-HSA domain III also hydrolyzed PNPA at an equivalent rate ($k_{obs} = [1.18 \pm 0.03] \times 10^{-4}$ s^{-1}), whereas almost no hydrolysis was observed in the absence of these synthetic domain III proteins. In the presence of (±)-ibuprofen (10 μM), lower rates of PNPA hydrolysis were obtained using L- and D-HSA domain III proteins, consistent with previously reported competition assays based on HSA-mediated PNPA hydrolysis [54]. These results suggested that the synthetic proteins hydrolyzed PNPA in a site II-dependent manner. The investigations of site II-dependent molecular recognition supported that the D enantiomer was correctly folded to form a mirror-image structure of bioactive native HSA domain III, which was consistent with their symmetrical CD spectra (Fig. 3.2a).

In summary, the author established a preparation protocol for synthetic HSA domain III involving convergent NCLs of seven peptide segments. The symmetrical structures of synthetic L- and D-HSA domain III proteins were confirmed using CD spectroscopy. Fluorescence-based binding assays using dansyl-Gly and SPR analysis using FcRn demonstrated that the ligand-binding properties of synthetic HSA domain III were comparable with those of native HSA. Preferential binding of dansyl-labeled L- and D-amino acids to site II of L- and D-HSA, respectively, indicated that HSA domain III recognized the chiral ligands in a stereoselective manner. Synthetic HSA domain III also hydrolyzed PNPA, confirming the protein had site-II-dependent esterase-like activity. These results supported that the synthetic D-HSA domain III would be applicable to screening studies using genetically-encoded L-peptide/protein libraries for exploring D-peptide- and D-protein-based ABMs. Notably, conjugation with D-peptide- and D-protein-based ABMs would prolong the serum half-lives of D-proteins-based therapeutic ligands without affecting their unique properties to facilitate their therapeutic applications.

3.1 Experimental Section

General Procedures of Peptide Synthesis. All reagents and solvents were purchased from Watanabe Chemical Industries, Ltd. (Hiroshima, Japan), Kokusan Chemical Industries, Ltd. (Kanagawa, Japan), Sigma-Aldrich JAPAN (Tokyo, Japan), Wako Pure Chemical Industries, Ltd. (Osaka, Japan), Tokyo Chemical Industry Co., Ltd. (Tokyo, Japan) or Nacalai Tesque, Inc. (Kyoto, Japan). For analytical HPLC, a Cosmosil 5C18-AR300 column (Nacalai Tesque, 4.6 × 250 mm) was employed with a linear gradient of CH_3CN containing 0.05% (v/v) TFA at a flow rate of 1 mL/min. The products were detected by UV absorbance at 220 nm. For preparative HPLC, a Cosmosil 5C18-AR300 preparative column (Nacalai Tesque, 20 × 250 mm) or a Cosmosil Protein-R preparative column (Nacalai Tesque, 20 × 150 mm) was employed with a linear gradient of CH_3CN containing 0.05% (v/v) TFA at flow rate of 8 mL/min. All peptides were characterized by ESI–MS [LCMS-2020 (Shimadzu)].

Fmoc-SPPS. Standard Fmoc-SPPS was performed using a CYCROSS automatic peptide synthesizer (DFC, Kyoto, Japan). The following side chain protected amino acids were employed: Arg(Pbf), Asn(Trt), Asp(O*t*-Bu), Cys(Trt), Gln(Trt), Glu(O*t*-Bu), His(Trt), Lys(Boc), Ser(*t*-Bu), Thr(*t*-Bu), Trp(Boc), Tyr(*t*-Bu). Fmoc-protected amino acids were coupled using OxymaPure (5 equiv)/DIC (5 equiv) in DMF for 60 min twice and Fmoc-deprotection was performed by 20% piperidine in DMF for 5 min twice.

Microwave-Assisted Fmoc-SPPS. Microwave-assisted peptide synthesis was performed on a Liberty BLUE (CEM Japan, Tokyo, Japan) peptide synthesizer. The following side chain protected amino acids were employed: Arg(Pbf), Asn(Trt), Asp(O*t*-Bu), Cys(Trt), Gln(Trt), Glu(O*t*-Bu), His(Trt), Lys(Boc), Ser(*t*-Bu), Thr(*t*-Bu), Trp(Boc), Tyr(*t*-Bu). Fmoc-protected amino acids were coupled using OxymaPure (5 equiv)/DIC (5 equiv) in DMF at 90 °C for 2 min, except for His(Trt). For coupling of Arg(Pbf), the reaction was repeated twice under the same conditions.

His(Trt) was coupled at 50 °C for 10 min. Fmoc-deprotection was performed by 20% piperidine in DMF for 1 min at 90 °C.

Synthesis of Fmoc-NHNH-Trt(2-Cl) resin. Fmoc-NHNH-Trt(2-Cl) resin was prepared according to the previously reported procedure [56]. A solution of Fmoc-$NHNH_2$ (321 mg, 1.24 mmol) and DIEA (1.06 mL, 6.06 mmol) in dry DMF/CH_2Cl_2 (4:1) was reacted with Trt(2-Cl) chloride resin (2.80 g, 3.03 mmol) for 3 h at room temperature.

[Cys(Acm)^{392}]-L-HSA$^{380–405}$ MPAA thioester (L-1). Asn405 was manually loaded on Fmoc-NHNH-Trt(2-Cl) resin (0.40 g, 0.10 mmol) by Fmoc-Asn(Trt)-OH (0.30 g, 0.50 mmol), OxymaPure (71 mg, 0.50 mmol), and DIC (78 μL, 0.50 mmol) in DMF. The remaining peptide sequence was constructed using a microwave-assisted peptide synthesizer. Fmoc-Cys(Acm)-OH was employed for the coupling of Cys392. The resin was then treated with TFA/H_2O/*m*-cresol/thioanisole/EDT (80:5:5:5:5) for 2 h at room temperature. After removal of the resin by filtration, the filtrate was poured into ice-cold dry Et_2O. The precipitate was dissolved in MPAA suspension (200 mM MPAA, 6 M guanidine; 5.0 mL). Then, acac (51 μL, 0.50 mmol) was added and the reaction was continued for 2 h at room temperature. After H_2O/CH_3CN (1:1) containing 0.05% (v/v) TFA (3.3 mL) was added, the mixture was purified by preparative HPLC (Cosmosil Protein-R column, a linear gradient of 25–55% CH_3CN in H_2O containing 0.05% (v/v) TFA over 90 min) to provide L-**1** (54 mg, 15% yield from resin). MS(ESI) *m/z*, calcd for $C_{152}H_{228}N_{36}O_{47}S_2$: 3373.60; observed: $[M + 2H]^{2+}$ *m/z* = 1689.00, $[M + 3H]^{3+}$ *m/z* = 1126.25.

[Cys^{406}]-L-HSA$^{406–436}$ (L-2). Lys436 was manually loaded on Fmoc-NHNH-Trt(2-Cl) resin (0.25 g, 0.10 mmol) by Fmoc-Lys(Boc)-OH (0.23 g, 0.50 mmol), OxymaPure (71 mg, 0.50 mmol), and DIC (78 μL, 0.50 mmol) in DMF. The remaining peptide sequence was constructed by using a microwave-assisted peptide synthesizer. The resin was then treated with TFA/H_2O/*m*-cresol/thioanisole/EDT (80:5:5:5:5) for 2 h at room temperature. After removal of the resin by filtration, the filtrate was poured into ice-cold dry Et_2O. The precipitate was purified by preparative HPLC (Cosmosil 5C18-AR300 column, a linear gradient of 20–50% CH_3CN in H_2O containing 0.05% (v/v) TFA over 90 min) to provide L-**2** (87.7 mg, 20% yield from resin). MS(ESI) *m/z*, calcd for $C_{151}H_{265}N_{45}O_{42}S$: 3412.97; observed: $[M + 2H]^{2+}$ *m/z* = 1708.40, $[M + 3H]^{3+}$ *m/z* = 1139.10, $[M + 4H]^{4+}$ *m/z* = 854.50.

L-HSA$^{437–460}$ (L-3). By the same protocol for the synthesis of peptide L-**2**, peptide L-**3** was synthesized (26.3 mg, 15% yield) from Fmoc-NHNH-Trt(2-Cl) resin (0.14 g, 0.050 mmol). MS(ESI) *m/z*, calcd for $C_{117}H_{193}N_{35}O_{34}S_4$: 2760.33; observed: $[M + 2H]^{2+}$ *m/z* = 1382.00, $[M + 3H]^{3+}$ *m/z* = 921.75, $[M + 4H]^{4+}$ *m/z* = 691.45.

[Cys(Acm)^{392}]-L-HSA$^{380–436}$ (L-4). Peptides L-**1** (51.6 mg) and L-**2** (47.3 mg) were dissolved in 1,2,4-triazole ligation buffer [2.5 M 1,2,4-triazole, 30 mM TCEP, 6 M guanidine, 100 mM phosphate (pH 7.2); 5.5 mL], and the reaction mixture was incubated for 1 h at 37 °C. Then, desulfurization buffer [450 mM TCEP, 80 mM MESNa, 6 M guanidine, 100 mM phosphate (pH 4.5); 5.5 mL] and VA-044 (89 mg) were added, and the pH was adjusted to 6.5. The reaction mixture was further incubated for 2 h at 37 °C. The product L-**4** was purified by preparative HPLC (Cosmosil 5C18-AR300 column, a linear gradient of 27–57% CH_3CN in H_2O containing 0.05% (v/v) TFA over 90 min). The remaining intermediate ([Cys(Acm)^{392}, Cys^{406}]-L-HSA$^{380–436}$) was also isolated and applied to desulfurization condition [20 mM VA-044, 250 mM TCEP, 100 mM MESNa, 6 M guanidine, 100 mM phosphate (pH 6.5)] to provide L-**4** (totally 40.5 mg, 48% yield). MS(ESI) *m/z*, calcd for $C_{295}H_{486}N_{81}O_{87}S$: 6587.58; observed: $[M + 4H]^{4+}$ *m/z* = 1648.60, $[M + 5H]^{5+}$ *m/z* = 1319.10, $[M + 6H]^{6+}$ *m/z* = 1099.40, $[M + 7H]^{7+}$ *m/z* = 942.50.

L-HSA$^{380–436}$ (L-5). Peptide L-**4** (21.7 mg) was dissolved in deprotection cocktail [AgOTf (27 mg), anisole (10 μL), TFA (0.50 mL)], and the reaction mixture was incubated for 1.5 h on ice. The product was poured into ice-cold Et_2O, and the precipitate was collected by centrifugation. The precipitate was washed by reduction buffer [10 mM DTT, 6 M guanidine PBS (pH 7.4)], and the supernatant was purified by preparative HPLC (Cosmosil 5C18-AR300 column, a linear gradient of 27–57% CH_3CN in H_2O containing 0.05% (v/v) TFA over 90 min) to provide L-**5** (14.8 mg, 69%). MS(ESI) *m/z*, calcd for $C_{292}H_{481}N_{80}O_{86}S$: 6516.54; observed: $[M + 4H]^{4+}$ *m/z* = 1630.50, $[M + 5H]^{5+}$ *m/z* = 1304.55, $[M + 6H]^{6+}$ *m/z* = 1087.25, $[M + 7H]^{7+}$ *m/z* = 932.05.

L-HSA$^{380–436}$ MPAA thioester (L-6). Acac (3.4 μL) was added to a solution of L-**5** (25.5 mg) in MPAA suspension (1.7 mL) and the reaction was continued for 1.5 h at room temperature. After H_2O/CH_3CN (1:1) containing 0.05% (v/v) TFA (0.62 mL) was added, the mixture was purified by preparative HPLC (Cosmosil Protein-R column, a linear gradient of 28–58% CH_3CN in H_2O containing 0.05% (v/v) TFA over 90 min) to provide L-**6** (19.6 mg, 77% yield). MS(ESI) *m/z*, calcd for $C_{300}H_{485}N_{78}O_{88}S_2$: 6652.53; observed: $[M + 4H]^{4+}$ *m/z* = 1664.85, $[M + 5H]^{5+}$ *m/z* = 1332.05, $[M + 6H]^{6+}$ *m/z* = 1110.20, $[M + 7H]^{7+}$ *m/z* = 951.75.

L-HSA$^{380–460}$ (L-7). Peptides L-**6** (19.6 mg) and L-**3** (12.6 mg) were dissolved in MPAA ligation buffer [100 mM MPAA, 50 mM TCEP, 6 M guanidine, 200 mM phosphate (pH 7.0); 2.5 mL], and the reaction mixture was incubated for 1 h at 37 °C. The reaction mixture was purified by preparative HPLC (Cosmosil Protein-R column, a linear gradient of 28–58% CH_3CN in H_2O containing 0.05% (v/v) TFA over 90 min) to provide L-**7** (20.7 mg, 75% yield). MS(ESI) *m/z*, calcd for $C_{409}H_{670}N_{113}O_{120}S_5$: 9244.84; observed: $[M + 5H]^{5+}$ *m/z* = 1850.70, $[M + 6H]^{6+}$ *m/z* = 1542.45, $[M + 7H]^{7+}$ *m/z* = 1322.10, $[M + 8H]^{8+}$ *m/z* = 1157.00, $[M + 9H]^{9+}$ *m/z* = 1028.50.

L-HSA$^{380–460}$ MPAA thioester (L-8). By the same protocol for the synthesis of peptide L-**6**, peptide L-**8** was synthesized (14.9 mg, 72% yield) from L-**7** (20.7 mg). MS(ESI) *m/z*, calcd for $C_{417}H_{674}N_{111}O_{122}S_6$: 9380.83; observed: $[M + 5H]^{5+}$ *m/z* = 1878.05, $[M + 6H]^{6+}$ *m/z* = 1565.10, $[M + 7H]^{7+}$ *m/z* = 1341.75, $[M + 8H]^{8+}$ *m/z* = 1174.10, $[M + 9H]^{9+}$ *m/z* = 1043.75.

[Thz[461]]-L-HSA$^{461–486}$ MPAA thioester (L-9). By the same protocol for the synthesis of peptide L-**1**, peptide L-**9** was synthesized (50.8 mg, 13% yield) from Fmoc-NHNH-Trt(2-Cl) resin (0.16 g, 0.10 mmol). The reaction for thioester formation was conducted for 18 h. MS(ESI) *m/z*, calcd for $C_{132}H_{218}N_{40}O_{40}S_4$: 3131.51; observed: $[M + 3H]^{3+}$ *m/z* = 1045.40, $[M + 4H]^{4+}$ *m/z* = 784.30.

L-HSA$^{487–513}$ (L-10). By the same protocol for the synthesis of peptide L-**2**, peptide L-**10** was synthesized (29.0 mg, 8% yield) from Fmoc-NHNH-Trt(2-Cl) resin (0.24 g, 0.050 mmol). MS(ESI) *m/z*, calcd for $C_{143}H_{205}N_{33}O_{45}S$: 3136.45; observed: $[M + 2H]^{2+}$ *m/z* = 1570.00, $[M + 3H]^{3+}$ *m/z* = 1047.00.

[Thz[514]]-L-HSA$^{514–557}$ (L-11). Poly-arginine tag was manually loaded on SAL resin (0.48 mmol/g, 0.21 g, 0.10 mmol) by Fmoc-Arg(Pbf)-OH (0.16 g, 0.25 mmol), OxymaPure (36 mg, 0.25 mmol) and DIC (39 μL, 0.25 mmol) in DMF for 2 h each step. MeDbz linker was then loaded on the resin by Fmoc-MeDbz-OH (97 mg, 0.25 mmol), HATU (95 mg, 0.25 mmol) and DIEA (87 μL, 0.50 mmol) in DMF for 3 h. The following Lys557 was coupled by Fmoc-Lys(Boc)-OH (0.23 g, 0.50 mmol), HATU (0.19 g, 0.50 mmol) and DIEA (0.17 mL, 1.0 mmol) in DMF for 2 h. The remaining peptide sequence was constructed using an automatic peptide synthesizer. Boc-Thz-OH was employed for coupling of the N-terminal Thz514. After the solid-phase peptide synthesis, a solution of *p*-nitrophenyl chloroformate in CH_2Cl_2 (90 mg/mL, 3 mL) was added to the protected peptide resin, and the reaction was continued for 2 h at room temperature. Subsequently, the resin was treated with 1 M DIEA in DMF (3 mL) for 2 h at 37 °C twice. The resin was then treated with TFA/H_2O/*m*-cresol/thioanisole/EDT (80:5:5:5:5) for 2 h at room temperature. After removal of the resin by filtration, the filtrate was poured into ice-cold dry Et_2O. The precipitate was purified by preparative HPLC (Cosmosil 5C18-AR300 column, a linear gradient of 22–52% CH_3CN in H_2O containing 0.05% (v/v) TFA over 90 min) to afford L-**11** (39.1 mg, 6% from resin). MS(ESI) *m/z*, calcd for $C_{258}H_{434}N_{80}O_{71}S_2$: 5853.23; observed: $[M + 4H]^{4+}$ *m/z* = 1468.20, $[M + 5H]^{5+}$ *m/z* = 1174.75, $[M + 6H]^{6+}$ *m/z* = 979.15.

L-HSA$^{558–585}$ (L-12). (+)-Biotin tag was manually loaded on SAL resin (0.47 mmol/g, 0.21 g, 0.10 mmol) by Fmoc-Lys(biotin)-OH (0.15 g, 0.25 mmol), OxymaPure (36 mg, 0.25 mmol) and DIC (39 μL, 0.25 mmol) in DMF/NMP (1:1) for 2 h. PEG linker was then loaded on the resin by Fmoc-NH-PEG$_2$-DGA-OH (140 mg, 0.25 mmol), OxymaPure (36 mg, 0.25 mmol) and DIC (39 μL, 0.25 mmol) in DMF for 2 h each step. The remaining peptide sequence was constructed using an automatic peptide synthesizer. The resin was then treated with TFA/H_2O/*m*-cresol/thioanisole/EDT (80:5:5:5:5) for 2 h at room temperature. After removal of the resin by filtration, the filtrate was poured into ice-cold dry Et_2O. The precipitate was purified by preparative HPLC (Cosmosil 5C18-AR300 column, a linear gradient of 24–54% CH_3CN in H_2O containing 0.05% (v/v) TFA over 90 min) to afford L-**12** (53.5 mg, 12% from resin). MS(ESI) *m/z*, calcd for $C_{166}H_{282}N_{42}O_{56}S_4$: 3887.94; observed: $[M + 3H]^{3+}$ *m/z* = 1297.65, $[M + 4H]^{4+}$ *m/z* = 973.45, $[M + 5H]^{5+}$ *m/z* = 779.00.

[Thz^{461}]-L-HSA$^{461–513}$ (L-13). Peptides L-**9** (70.4 mg) and L-**10** (40.8 mg) were dissolved in MPAA ligation buffer with increased TCEP concentration [100 mM MPAA, 100 mM TCEP, 6 M guanidine, 200 mM phosphate (pH 7.1); 5.6 mL], and the reaction mixture was incubated for 18 h at 37 °C. The reaction mixture was purified by preparative HPLC (Cosmosil Protein-R column, a linear gradient of 22–52% CH_3CN in H_2O containing 0.05% (v/v) TFA over 90 min) to afford L-**13** (41.0 mg, 50%). MS(ESI) *m/z*, calcd for $C_{267}H_{415}N_{73}O_{83}S_4$: 6099.94; observed: $[M + 4H]^{4+}$ *m/z* = 1526.90, $[M + 5H]^{5+}$ *m/z* = 1221.65, $[M + 6H]^{6+}$ *m/z* = 1018.20, $[M + 7H]^{7+}$ *m/z* = 872.90.

[Thz^{461}]-L-HSA$^{461–513}$ MPAA thioester (L-14). By the same protocol for the synthesis of peptide L-**6**, peptide L-**14** was synthesized (8.2 mg, 75% yield) from L-**13** (10.8 mg). MS(ESI) *m/z*, calcd for $C_{275}H_{419}N_{71}O_{85}S_5$: 6235.93; observed: $[M + 4H]^{4+}$ *m/z* = 1560.90, $[M + 5H]^{5+}$ *m/z* = 1248.90, $[M + 6H]^{6+}$ *m/z* = 1040.80, $[M + 7H]^{7+}$ *m/z* = 892.35.

L-HSA$^{514–585}$ (L-15). Peptides L-**11** (24.2 mg) and L-**12** (11.2 mg) were dissolved in MPAA ligation buffer (2.5 mL), and the reaction mixture was incubated for 1 h at 37 °C. Then, 2 M $MeONH_2$ (0.28 mL) was added, and the reaction mixture was incubated for 2 h at 37 °C. The reaction mixture was purified by preparative HPLC (Cosmosil Protein-R column, a linear gradient of 30–60% CH_3CN in H_2O containing 0.05% (v/v) TFA over 90 min) to provide L-**15** (11.8 mg, 44% yield). MS(ESI) *m/z*, calcd for $C_{391}H_{659}N_{103}O_{121}S_6$: 8925.69; observed: $[M + 5H]^{5+}$ *m/z* = 1787.20, $[M + 6H]^{6+}$ *m/z* = 1489.40, $[M + 7H]^{7+}$ *m/z* = 1276.80, $[M + 8H]^{8+}$ *m/z* = 1117.35, $[M + 9H]^{9+}$ *m/z* = 993.25.

L-HSA$^{461–585}$ (L-16). Peptides L-**14** (4.1 mg) and L-**15** (4.4 mg) were dissolved in MPAA ligation buffer with increased TCEP concentration (0.41 mL), and the reaction mixture was incubated for 10 h at 37 °C. Then, L-**14** (2.3 mg) in MPAA ligation buffer (0.41 mL) was added, and the reaction mixture was incubated for 5 h at 37 °C. Subsequently, 2 M $MeONH_2$ (91 μL) and TCEP (12 mg) were added to the reaction mixture and the reaction was continued for 2 h at 37 °C. The reaction mixture was purified by preparative HPLC (Cosmosil Protein-R column, a linear gradient of 30–60% CH_3CN in H_2O containing 0.05% (v/v) TFA over 90 min) to provide L-**16** (3.2 mg, 44% yield). MS(ESI) *m/z*, calcd for $C_{657}H_{1070}N_{174}O_{204}S_{10}$: 14981.59; observed: $[M + 9H]^{9+}$ *m/z* = 1666.60, $[M + 10H]^{10+}$ *m/z* = 1500.05, $[M + 11H]^{11+}$ *m/z* = 1363.80, $[M + 12H]^{12+}$ *m/z* = 1250.20, $[M + 13H]^{13+}$ *m/z* = 1154.55, $[M + 14H]^{14+}$ *m/z* = 1071.70, $[M + 15H]^{15+}$ *m/z* = 1000.65.

L-HSA$^{380–585}$ (L-17). Peptides L-**8** (7.8 mg) and L-**16** (8.7 mg) were dissolved in MPAA ligation buffer (2.0 mL), and the reaction mixture was incubated for 1.5 h at 37 °C. The reaction mixture was purified by preparative HPLC (Cosmosil 5C18-AR300 column, a linear gradient of 32–62% CH_3CN in H_2O containing 0.05% (v/v) TFA over 90 min) to provide L-**17** (4.9 mg, 35% yield). MS(ESI) *m/z*, calcd for $C_{1066}H_{1736}N_{285}O_{324}S_{15}$: 24194.39; observed: $[M + 18H]^{18+}$ *m/z* = 1345.45, $[M + 19H]^{19+}$ *m/z* = 1274.55, $[M + 20H]^{20+}$ *m/z* = 1210.95, $[M + 21H]^{21+}$ *m/z* = 1153.55,

$[M + 22H]^{22+}$ *m/z* = 1101.00, $[M + 23H]^{23+}$ *m/z* = 1053.20, $[M + 24H]^{24+}$ *m/z* = 1009.30, $[M + 25H]^{25+}$ *m/z* = 968.95.

[Cys(Acm)^{392}]-D-HSA$^{380–405}$ MPAA thioester (D-1). By the identical procedure for the synthesis of peptide L-**1**, peptide D-**1** was synthesized (33.3 mg, 9% yield) from Fmoc-NHNH-Trt(2-Cl) resin (0.25 g, 0.10 mmol). MS(ESI) *m/z*, calcd for $C_{152}H_{228}N_{36}O_{47}S_2$: 3373.60; observed: $[M + 2H]^{2+}$ *m/z* = 1688.50, $[M + 3H]^{3+}$ *m/z* = 1125.95.

[Cys^{406}]-D-HSA$^{406–436}$ (D-2). By the identical procedure for the synthesis of peptide L-**2**, peptide D-**2** was synthesized (54.1 mg, 13% yield) from Fmoc-NHNH-Trt(2-Cl) resin (0.25 g, 0.10 mmol). MS(ESI) *m/z*, calcd for $C_{151}H_{265}N_{45}O_{42}S$: 3412.97; observed: $[M + 2H]^{2+}$ *m/z* = 1708.65, $[M + 3H]^{3+}$ *m/z* = 1139.30, $[M + 4H]^{4+}$ *m/z* = 854.75.

D-HSA$^{437–460}$ (D-3). By the identical procedure for the synthesis of peptide L-**3**, peptide D-**3** was synthesized (25.9 mg, 15% yield) from Fmoc-NHNH-Trt(2-Cl) resin (0.12 g, 0.050 mmol). MS(ESI) *m/z*, calcd for $C_{117}H_{193}N_{35}O_{34}S_4$: 2760.33; observed: $[M + 2H]^{2+}$ *m/z* = 1381.95, $[M + 3H]^{3+}$ *m/z* = 921.65, $[M + 4H]^{4+}$ *m/z* = 691.50.

[Cys(Acm)^{392}]-D-HSA$^{380–436}$ (D-4). By the identical procedure for the synthesis of peptide L-**4**, peptide D-**4** was synthesized (47.2 mg, 49% yield) from D-**1** (54.2 mg) and D-**2** (54.1 mg). MS(ESI) *m/z*, calcd for $C_{295}H_{486}N_{81}O_{87}S$: 6587.58; observed: $[M + 4H]^{4+}$ *m/z* = 1648.60, $[M + 5H]^{5+}$ *m/z* = 1319.15, $[M + 6H]^{6+}$ *m/z* = 1099.30, $[M + 7H]^{7+}$ *m/z* = 942.45.

D-HSA$^{380–436}$ (D-5). By the identical procedure for the synthesis of peptide L-**5**, peptide D-**5** was synthesized (35.3 mg, 75% yield) from D-**4** (47.2 mg). MS(ESI) *m/z*, calcd for $C_{292}H_{481}N_{80}O_{86}S$: 6516.54; observed: $[M + 4H]^{4+}$ *m/z* = 1630.80, $[M + 5H]^{5+}$ *m/z* = 1304.65, $[M + 6H]^{6+}$ *m/z* = 1087.45, $[M + 7H]^{7+}$ *m/z* = 932.30.

D-HSA$^{380–436}$ MPAA thioester (D-6). By the identical procedure for the synthesis of peptide L-**6**, peptide D-**6** was synthesized (24.3 mg, 71% yield) from D-**5** (34.3 mg). MS(ESI) *m/z*, calcd for $C_{300}H_{485}N_{78}O_{88}S_2$: 6652.53; observed: $[M + 4H]^{4+}$ *m/z* = 1665.00, $[M + 5H]^{5+}$ *m/z* = 1332.10, $[M + 6H]^{6+}$ *m/z* = 1110.25.

D-HSA$^{380–460}$ (D-7). By the identical procedure for the synthesis of peptide L-**7**, peptide D-**7** was synthesized (23.3 mg, 68% yield) from D-**6** (24.3 mg) and D-**3** (14.0 mg). MS(ESI) *m/z*, calcd for $C_{409}H_{670}N_{113}O_{120}S_5$: 9244.84; observed: $[M + 5H]^{5+}$ *m/z* = 1851.05, $[M + 6H]^{6+}$ *m/z* = 1542.70, $[M + 7H]^{7+}$ *m/z* = 1322.45, $[M + 8H]^{8+}$ *m/z* = 1157.15, $[M + 9H]^{9+}$ *m/z* = 1028.70.

D-HSA$^{380–460}$ MPAA ester (D-8). By the identical procedure for the synthesis of peptide L-**8**, peptide D-**8** was synthesized (19.0 mg, 81% yield) from D-**7** (23.3 mg). MS(ESI) *m/z*, calcd for $C_{417}H_{674}N_{111}O_{122}S_6$: 9380.83; observed: $[M + 5H]^{5+}$ *m/z* = 1878.10, $[M + 6H]^{6+}$ *m/z* = 1565.05, $[M + 7H]^{7+}$ *m/z* = 1341.75, $[M + 8H]^{8+}$ *m/z* = 1174.00, $[M + 9H]^{9+}$ *m/z* = 1043.60.

[Thz^{461}]-D-HSA$^{461–486}$ MPAA thioester (D-9). By the identical procedure for the synthesis of peptide L-**9**, peptide D-**9** was synthesized (79.4 mg, 10% yield) from Fmoc-NHNH-Trt(2-Cl) resin (0.50 g, 0.20 mmol). MS(ESI) *m/z*, calcd for $C_{132}H_{218}N_{40}O_{40}S_4$: 3131.51; observed: $[M + 3H]^{3+}$ *m/z* = 1045.40, $[M + 4H]^{4+}$ *m/z* = 784.30.

D-HSA$^{487–513}$ (D-10). By the identical procedure for the synthesis of peptide L-**10**, peptide D-**10** was synthesized (17.3 mg, 5% yield) from Fmoc-NHNH-Trt(2-Cl) resin (0.25 g, 0.10 mmol). MS(ESI) *m/z*, calcd for $C_{143}H_{205}N_{33}O_{45}S$: 3136.45; observed: $[M + 2H]^{2+}$ *m/z* = 1570.35, $[M + 3H]^{3+}$ *m/z* = 1047.05.

[Thz^{514}]-D-HSA$^{514–557}$ (D-11). By the identical procedure for the synthesis of peptide L-**11**, peptide D-**11** was synthesized (42.3 mg, 5% yield) from Fmoc-NH-SAL resin (0.25 g, 0.12 mmol). MS(ESI) *m/z*, calcd for $C_{258}H_{434}N_{80}O_{71}S_2$: 5853.23; observed: $[M + 4H]^{4+}$ *m/z* = 1468.05, $[M + 5H]^{5+}$ *m/z* = 1174.50.

D-HSA$^{558–585}$ (D-12). Poly-histidine tag (L-form) was loaded on TGR resin (0.25 mmol, 0.40 g, 0.10 mmol) using a microwave-assisted peptide synthesizer. (+)-Biotin tag was manually loaded by Fmoc-Lys(biotin)-OH (0.15 g, 0.25 mmol), OxymaPure (36 mg, 0.25 mmol) and DIC (39 μL, 0.25 mmol) in DMF/NMP (1:1) for 2 h. The remaining peptide sequence was constructed by the identical procedure for the synthesis of peptide L-**12**. The resin was then treated with TFA/H_2O/*m*-cresol/thioanisole/EDT (80:5:5:5:5) for 24 h at room temperature. After removal of the resin by filtration, the filtrate was poured into ice-cold dry Et_2O. The precipitate was purified by preparative HPLC (Cosmosil 5C18-AR300 column, a linear gradient of 23–53% CH_3CN in H_2O containing 0.05% (v/v) TFA over 90 min) to afford D-**12** (38.9 mg, 7% yield). MS(ESI) *m/z*, calcd for $C_{202}H_{324}N_{60}O_{62}S_4$: 4710.29; observed: $[M + 3H]^{3+}$ *m/z* = 1572.05, $[M + 4H]^{4+}$ *m/z* = 1179.20, $[M + 5H]^{5+}$ *m/z* = 943.60.

[Thz^{461}]-D-HSA$^{461–513}$ (D-13). By the identical procedure for the synthesis of peptide L-**13**, peptide D-**13** was synthesized (32.5 mg, 51% yield) from D-**9** (45.5 mg) and D-**10** (31.9 mg). MS(ESI) *m/z*, calcd for $C_{267}H_{415}N_{73}O_{83}S_4$: 6099.94; observed: $[M + 4H]^{4+}$ *m/z* = 1526.70, $[M + 5H]^{5+}$ *m/z* = 1221.60, $[M + 6H]^{6+}$ *m/z* = 1018.10, $[M + 7H]^{7+}$ *m/z* = 872.75.

[Thz^{461}]-D-HSA$^{461–513}$ MPAA thioester (D-14). By the identical procedure for the synthesis of peptide L-**14**, peptide D-**14** was synthesized (27.0 mg, 82% yield) from D-**13** (32.5 mg). MS(ESI) *m/z*, calcd for $C_{275}H_{419}N_{71}O_{85}S_5$: 6235.93; observed: $[M + 4H]^{4+}$ *m/z* = 1560.85, $[M + 5H]^{5+}$ *m/z* = 1248.70, $[M + 6H]^{6+}$ *m/z* = 1040.95, $[M + 7H]^{7+}$ *m/z* = 892.15.

D-HSA$^{514–585}$ (D-15). By the identical procedure for the synthesis of peptide L-**15**, peptide D-**15** was synthesized (17.7 mg, 34% yield) from D-**11** (37.5 mg) and D-**12** (25.3 mg). MS(ESI) *m/z*, calcd for $C_{427}H_{701}N_{121}O_{127}S_6$: 9748.04; observed: $[M + 6H]^{6+}$ *m/z* = 1626.60, $[M + 7H]^{7+}$ *m/z* = 1394.35, $[M + 8H]^{8+}$ *m/z* = 1220.15, $[M + 9H]^{9+}$ *m/z* = 1084.65, $[M + 10H]^{10+}$ *m/z* = 976.25.

D-HSA$^{461–585}$ (D-16). By the identical procedure for the synthesis of peptide L-**16**, peptide D-**16** was synthesized (6.0 mg, 51% yield) from D-**14** (6.8 mg + 3.0 mg) and D-**15** (7.5 mg). MS(ESI) *m/z*, calcd for $C_{693}H_{1112}N_{192}O_{210}S_{10}$: 15803.94; observed: $[M + 12H]^{12+}$ *m/z* = 1318.75, $[M + 13H]^{13+}$ *m/z* = 1217.45 $[M + 14H]^{14+}$ *m/z* = 1130.55, $[M + 15H]^{15+}$ *m/z* = 1055.50, $[M + 16H]^{16+}$ *m/z* = 989.35, $[M + 17H]^{17+}$ *m/z* = 931.35.

D-HSA$^{380–585}$ (D-17). By the identical procedure for the synthesis of peptide L-**17**, peptide D-**17** was synthesized (4.3 mg, 46% yield) from D-**8** (10.9 mg) and D-**16** (6.0 mg). MS(ESI) *m/z*, calcd for $C_{1102}H_{1778}N_{303}O_{330}S_{15}$: 25016.75; observed: $[M + 18H]^{18+}$ *m/z* = 1391.75, $[M + 19H]^{19+}$ *m/z* = 1318.35, $[M + 20H]^{20+}$ *m/z* = 1252.45, $[M + 21H]^{21+}$ *m/z* = 1192.95, $[M + 22H]^{22+}$ *m/z* = 1138.70, $[M + 23H]^{23+}$ *m/z* = 1089.15, $[M + 24H]^{24+}$ *m/z* = 1043.70.

Refolding of HSA Domain III. The lyophilized protein was dissolved in urea solution [50 mM dithiothreitol, 1 mM ethylenediaminetetraacetic acid (EDTA), 8 M urea, 50 mM carbonate (pH 10); 2 mg/mL], and incubated for 30 min at 37 °C. The denatured protein was diluted with an equal volume of carbonate buffer [1 mM EDTA, 50 mM carbonate (pH 10)] and then subjected to repeated dialysis against a 100-fold volume of carbonate buffer [1 mM EDTA, 50 mM carbonate (pH 10)] using a Slide-A-Lyzer G2 dialysis cassette with a 3.5 kDa cutoff (Thermo Fisher Scientific, MA, USA), first for 2 h and then 5 h at room temperature. The protein solution was further dialyzed against a 100-fold volume of oxidation buffer [1 mM cystamine, 1 mM EDTA, 50 mM carbonate (pH 10)] for 18 h at room temperature. The folded protein was purified in phosphate-buffered saline (PBS; pH 7.4) using preparative SEC with an ENrich SEC70 10 × 300 column (Bio-Rad Laboratories, CA, USA) and a flow rate of 0.8 mL/min.

Confirmation of Disulfide Bond Formations. Solutions of refolded L- or D-HSA domain III in PBS were desalted using dialysis and then lyophilized. L- and D-HSA domain III proteins were dissolved in denaturation buffer [6 M guanidine, 20 mM Tris (pH 8)] at concentrations of ≈0.3 mg/mL. To the protein solutions (40 μL), reduction buffer [0.1 M TCEP, 20 mM Tris (pH 8); 2 μL] or control buffer [20 mM Tris (pH 8); 2 μL] was added, and the mixtures were incubated for 30 min at room temperature. Then, alkylation buffer [0.1 M iodoacetamide, 20 mM Tris (pH 8); 4 μL] was added, and the reactions were continued for 1 h at room temperature. The products were analyzed using LC–MS.

Preparation of Recombinant HSA Domain III. HSA domain III (residues 381–585) was recombinantly expressed following a previously reported procedure [55]. Briefly, the gene segment coding for HSA domain III was amplified using a polymerase chain reaction with a cDNA clone of HSA as a template. The 5′ sense primer encoded an *Xho*I restriction site. The 3′ antisense primer was designed to incorporate an *Eco*RI site and add a stop codon to terminate the translation. The sequence of the 5′ primer was 5′-GGACTAGTCTCGAGAAAAGAGTGGAAGAGCCTCAG-3′. The sequence of the 3′ primer was 5′-CGCGAATTCTTATAAGCCTAAGGCAGC-3′. The amplified gene segments were ligated into a pPIC9 vector, which contains an α-factor secretion signal sequence. The resulting vector was introduced into the yeast species *Pichia pastoris* (strain GS115). The secreted domain III protein was isolated from the growth medium using sequential HPLC purifications with a Blue Sepharose 6 Fast Flow column (Cytiva, Tokyo, Japan) and a Phenyl HP column (Cytiva). HSA domain III was further purified using SEC (ENrich SEC 70 10 × 300 column) in PBS (pH 7.4).

CD Spectra Measurement. Proteins in PBS (pH 7.4) were used for measurement. CD spectrum was recorded on a J-1500 circular dichroism spectrometer (JASCO) at 25 °C. Full-length HSA (Nacalai Tesque) was used for CD measurement and following assays.

SPR Analysis. Binding of synthetic HSA domain III or full-length HSA to FcRn was analyzed using a Biacore X-100 system (Cytiva). Recombinant human FcRn protein (R & D Systems, MN, USA; cat# 8639-FC) was immobilized on a CM5 sensor chip (Cytiva) using an amine coupling kit with HBS-EP$^+$ buffer (Cytiva) as a running buffer (≈1100 RU). The analyte protein solution was injected at a rate of 30 μL/min into the flow system in running buffer [200 mM phosphate, 150 mM NaCl, 0.005% Tween-20 (pH 5.5)] at 25 °C. Association of all analytes was evaluated for 90 s, then dissociation was monitored for 600 s. HBS-EP$^+$ buffer was used for regeneration. Binding affinities were determined from triplicate assays by fitting a 1:1 binding model.

Analysis of Dansyl-Gly Binding to HSA Domain III. Fluorescent probe-based assays were carried out according to the previous investigations [46]. 96-Well half-area black plates (Corning Inc., NY, USA; cat# 3686) were employed for the assays. A solution (25 μL/well) of dansyl-Gly (20 μM) in PBS containing 1% (v/v) DMSO was added to protein (0–8 μM) in PBS (25 μL/well) and incubated for 30 min at room temperature. Fluorescence (λ_{exc} = 365 nm, λ_{emi} = 480 nm) was measured for each well using an Infinite 200 Pro microplate reader (Tecan, Männedorf, Switzerland).

Synthesis of Dansyl Amino Acids. Dansyl-L-Phe [47], -L-Pro [47], -L-Val [48], -D-Phe [57], -D-Pro [58] and -D-Val [57] were synthesized according to the previous report [59]. Briefly, dansyl chloride (0.9 equiv) in acetone was added to the solution of amino acid (1.0 equiv) in saturated $NaHCO_3$ and the reaction mixture was stirred for 4–10 h at room temperature. After the removal of acetone by evaporation, 1 M HCl was added to the mixture to adjust the pH ≈3. The whole was extracted with EtOAc and the extract was washed with brine, and dried over Na_2SO_4. The filtrate was concentrated under reduced pressure and recrystallized from hexane/EtOAc. Dansyl-L-Pro and -D-Pro were isolated as salts of cyclohexylamine.

Binding of Chiral Dansyl Amino Acids to L- and D-HSA Domain III Proteins. The concentrations of dansyl amino acid solutions were equally adjusted on the basis of their absorbance at 325 nm. Solutions (25 μL/well) of individual dansyl amino acids (6 μM) in PBS containing 1% (v/v) DMSO were added to each of the protein solutions (6 μM) in PBS (25 μL/well) and incubated for 30 min at room temperature.

Fluorescence ($\lambda_{exc} = 365$ nm, $\lambda_{emi} = 480$ nm) was measured for each well using the Infinite 200 Pro microplate reader.

Competitive HSA Domain III Binding Assay Using Dansyl-Gly. Premixed solutions (25 μL/well) of dansyl-Gly (6 μM) and increasing concentrations of (±)-ibuprofen or (±)-warfarin (0–60 μM) in PBS containing 1% (v/v) DMSO were added to HSA domain III (6 μM) in PBS (25 μL/well) and incubated for 30 min at room temperature. Fluorescence ($\lambda_{exc} = 365$ nm, $\lambda_{emi} = 480$ nm) was measured for each well using the Infinite 200 Pro microplate reader.

Analysis of Esterase-like Activity of HSA Domain III. Enzyme assays were carried out using a UV–Vis-NIR spectrophotometer (UV-3600; Shimadzu, Kyoto, Japan). A premixed solution (2 μL) of PNPA (10 mM) and (±)-ibuprofen (0 or 1 mM) in CH_3CN was added to 200 μL protein (5 μM) in 100 mM phosphate buffer (pH 7.9). The UV absorbance at 400 nm was measured for 90 s, and k_{obs} was determined from linear regression analysis of a semi-logarithmic plot of PNPA concentration versus time. The concentration of PNPA at each time point was determined by subtracting the PNP concentration ($\varepsilon = 1.8 \times 10^4$ M^{-1} cm^{-1} at 400 nm) from the initial PNPA concentration.

References

1. Fanali G et al (2012) Human serum albumin: From bench to bedside. Mol Aspects Med 33:209–290
2. Rabbani G, Ahn SN (2019) Structure, enzymatic activities, glycation and therapeutic potential of human serum albumin: A natural cargo. Int J Biol Macromol 123:979–990
3. Sand KMK et al (2015) Unraveling the interaction between FcRn and albumin: Opportunities for design of albumin-based therapeutics. Front Immunol 5:682
4. Sugio S et al (1999) Crystal structure of human serum albumin at 2.5 Å resolution. Protein Eng 12:439–446
5. Ghuman J et al (2005) Structural basis of the drug-binding specificity of human serum albumin. J Mol Biol 353:38–52
6. Schmidt MM et al (2013) Crystal structure of an HSA/FcRn complex reveals recycling by competitive mimicry of HSA ligands at a pH-dependent hydrophobic interface. Structure 21:1966–1978
7. Dockal M et al (1999) The three recombinant domains of human serum albumin. J Biol Chem 274:29303–29310
8. Mao H et al (2000) Expression, refolding, and isotopic labeling of human albumin domains for NMR spectroscopy. Protein Expr Purif 20:492–499
9. Nilsen J et al (2018) Human and mouse albumin bind their respective neonatal Fc receptors differently. Sci Rep 8:14648
10. Tao H et al (2021) The development of human serum albumin-based drugs and relevant fusion proteins for cancer therapy. Int J Biol Macromol 187:24–34
11. Yardley DAJ (2013) *Nab*-Paclitaxel mechanisms of action and delivery. Control Release 170:365–372
12. Bak M et al (2020) Recombinant peptide production platform coupled with site-specific albumin conjugate enables a convenient production of long-acting therapeutic peptide. Pharmaceutics 12:364

13. Ullah A et al (2023) Human serum albumin binders: A piggyback ride for long-acting therapeutics. Drug Discov Today 28:103738
14. Tanaka Y (2023) Ozoralizumab: First Nanobody® therapeutic for rheumatoid arthritis. Expert Opin Biol Ther 23:579–587
15. Coppieters K et al (2006) Formatted anti-tumor necrosis factor α VHH proteins derived from camelids show superior potency and targeting to inflamed joints in a murine model of collagen-induced arthritis. Arthritis Rheum 54:1856–1866
16. Zorzi A et al (2019) Non-covalent albumin-binding ligands for extending the circulating half-life of small biotherapeutics. Med Chem Commun 10:1068–1081
17. Zorzi A et al (2017) Acylated heptapeptide binds albumin with high affinity and application as tag furnishes long-acting peptides. Nat Commun 8:16092
18. Steiner D et al (2017) Half-life extension using serum albumin-binding DARPin® domains. Protein Eng Des Sel 30:583–591
19. Jacobs SA et al (2015) Fusion to a highly stable consensus albumin binding domain allows for tunable pharmacokinetics. Protein Eng Des Sel 28:385–393
20. Lander AJ et al (2023) D-Peptide and D-protein technology: Recent advances, challenges, opportunities. ChemBioChem 24:e202200537
21. Dintzis HM et al (1993) A comparison of the immunogenicity of a pair of enantiomeric proteins. Proteins Struct Funct Genet 16:306–308
22. Uppalapati M et al (2016) A potent D-protein antagonist of VEGF-A is nonimmunogenic, metabolically stable, and longer-circulating in vivo. ACS Chem Biol 11:1058–1065
23. Marinec PS et al (2021) A non-immunogenic bivalent D-protein potently inhibits retinal vascularization and tumor growth. ACS Chem Biol 16:548–556
24. Schumacher TNM et al (1996) Identification of D-peptide ligands through mirror-image phage display. Science 271:1854–1857
25. Dawson PE et al (1994) Synthesis of proteins by native chemical ligation. Science 266:776–779
26. Hackeng TM et al (1999) Protein synthesis by native chemical ligation: Expanded scope by using straightforward methodology. Proc Natl Acad Sci USA 96:10068–10073
27. Fang GM et al (2011) Protein chemical synthesis by ligation of peptide hydrazides. Angew Chem Int Ed 50:7645–7649
28. Tian X et al (2016) Optimal peptide hydrazide ligation with C-terminus Asp, Asn, and Gln hydrazides. Tetrahedron Lett 57:4264–4267
29. Flood DT et al (2018) Leveraging the Knorr pyrazole synthesis for the facile generation of thioester surrogates for use in native chemical ligation. Angew Chem Int Ed 57:11634–11639
30. Sakamoto K et al (2017) 1,2,4-Triazole-aided native chemical ligation between peptide-*N*-acyl-*N*'-methyl-benzimidazolinone and cysteinyl peptide. Chem Commun 53:12236–12239
31. Wan Q, Danishefsky SJ (2007) Free-radical-based, specific desulfurization of cysteine: A powerful advance in the synthesis of polypeptides and glycopolypeptides. Angew Chem Int Ed 46:9248–9252
32. Hunter MJ, Komives EA (1995) Deprotection of *S*-acetamidomethyl cysteine-containing peptides by silver. Anal Biochem 228:173–177
33. Siman P et al (2013) Convergent chemical synthesis of histone H2B protein for the site-specific ubiquitination at Lys34. Angew Chem Int Ed 52:8059–8063
34. Blanco-Canosa JB et al (2015) Chemical protein synthesis using a second-generation *N*-acylurea linker for the preparation of peptide-thioester precursors. J Am Chem Soc 137:7197–7209
35. Raibaut L et al (2013) Bis(2-sulfanylethyl)amido peptides enable native chemical ligation at proline and minimize deletion side-product formation. Org Lett 15:5516–5519
36. Nakamura T et al (2014) Examination of native chemical ligation using peptidyl prolyl thioesters. Chem Commun 50:58–60
37. Burton SJ et al (1989) Refolding human serum albumin at relatively high protein concentration. Eur J Biochem 179:379–387
38. Johanson KO et al (1981) Refolding of bovine serum albumin and its proteolytic fragments. Regain of disulfide bonds, secondary structure, and ligand-binding ability. J Biol Chem 256:445–450

39. Ashraf S et al (2023) Unraveling the versality of human serum albumin—A comprehensive review of its biological significance and therapeutic potential. Curr Res Struct Biol 6:100114
40. Curry S (2009) Lessons from the crystallographic analysis of small molecule binding to human serum albumin. Drug Metab Pharmacokinet 24:342–357
41. Wang A et al (2020) Characterizing the interaction between methyl ferulate and human serum albumin by saturation transfer difference NMR. RSC Adv 10:32999–33009
42. Tramarin A et al (2019) New insights into the altered binding capacity of pharmaceutical-grade human serum albumin: Site-specific binding studies by induced circular dichroism spectroscopy. J Pharm Biomed Anal 162:171–178
43. Ge F et al (2011) Interaction between alizarin and human serum albumin by fluorescence spectroscopy. Anal Sci 27:79–84
44. Wang Y et al (2024) Development of human serum albumin fluorescent probes in detection, imaging, and disease therapy. J Phys Chem B 128:1121–1138
45. Er JC et al (2013) Fluorescent dye cocktail for multiple drug-site mapping on human serum albumin. ACS Comb Sci 15:452–457
46. Jithinraj TK et al (2023) Chiral 8-aminoBODIPY-based fluorescent probes with site selectivity for the quantitative detection of HSA in biological samples. Analyst 148:286–296
47. Sudlow G et al (1975) The characterization of two specific drug binding sites on human serum albumin. Mol Pharmacol 11:824–832
48. Sudlow G et al (1976) Further characterization of specific drug binding sites on human serum albumin. Mol Pharmacol 12:1052–1061
49. Bertozo LDC et al (2021) Binding of fluorescent dansyl amino acids in albumin: When access to the protein cavity is more important than the strength of binding. Dyes Pigm 188:109195
50. Peyrin E et al (1998) Peculiarities of dansyl amino acid enantioselectivity using human serum albumin as a chiral selector. J Chromatogr Sci 36:97–103
51. Yang F et al (2007) Effect of human serum albumin on drug metabolism: Structural evidence of esterase activity of human serum albumin. J Struct Biol 157:348–355
52. Lockridge O et al (2008) Pseudo-esterase activity of human albumin. Slow turnover on tyrosine 411 and stable acetylation of 82 residues including 59 lysines. J Biol Chem 283:22582–22590
53. Means GE, Bender ML (1975) Acetylation of human serum albumin by *p*-nitrophenyl acetate. Biochem 14:4989–4994
54. Tatsumi A et al (2016) Differences in esterase activity to aspirin and *p*-nitrophenyl acetate among human serum albumin preparations. Biol Pharm Bull 39:1364–1369
55. Matsushita S et al (2004) Functional analysis of recombinant human serum albumin domains for pharmaceutical applications. Pharm Res 21:1924–1932
56. Huang YC et al (2014) Facile synthesis of C-terminal peptide hydrazide and thioester of NY-ESO-1 (A39–A68) from an Fmoc-hydrazine 2-chlorotrityl chloride resin. Tetrahedron 70:2951–2955
57. Peyrin E et al (1997) Interactions between dansyl amino acids and human serum albumin using high-performance liquid chromatography: Mobile-phase pH and temperature considerations. Anal Chem 69:4979–4984
58. Wang F, Schwabacher AW (1999) Tuning of binding selectivity: Metal control of organic guest binding and allosteric perturbation of fluorescent metal sensor. J Org Chem 64:8922–8928
59. Sha C et al (2022) A dansyl-based fluorescent probe for turn-off and turn-on detection of Hg^{2+} in a full water system. Res Chem Intermed 48:5227–5241

Chapter 4
Development of Monobody Variants for Functional Mirror-Image Protein Therapeutics

Abstract Mirror-image proteins (D-proteins) are promising scaffolds for drug discovery because of their high proteolytic stability and low immunogenic properties. Facile and reproducible processes for the preparation of functional D-proteins are required for their application in therapeutic biologics. In this study, the author designed and synthesized novel monobody variants with two cysteine substitutions that facilitate the synthetic process via sequential native chemical ligations. The synthetic anti-GFP monobody in initial model study exhibited good binding affinity to the target enhanced green fluorescent protein. In vivo administration of the mirror-image anti-GFP monobody to mice induced no anti-drug antibody production. The structure–activity relationship study of anti-D-MCP-1 monobody revealed that cysteine substitutions at two appropriate positions in combination with cysteine-selective chemical stapling improved the thermal stability of the scaffold while keeping the high-affinity target binding. These results highlight that the synthetic D-monobody variants are functional protein scaffolds with low immunogenic properties.

Keywords Monobody · Protein scaffold · Chemical protein synthesis · Mirror-image protein · Immunogenicity · Monocyte chemoattractant protein 1 · Cysteine modification · Chemical stapling

Monobody is a non-immunoglobulin protein scaffold derived from the tenth type III domain of human fibronectin (FN3) [1]. The immunoglobulin domain-like β-sandwich structure of FN3 can hold several complementarity determining region (CDR)-like variable regions in the sequence. This rigid scaffold facilitates target binding with high affinity and selectivity by restricting conformational flexibility. Two types of monobodies were designed for screening variable regions (Fig. 4.1): three loops (BC, DE and FG loops) between β-strands contain variable regions in a loop-only library, and two β-strands (C and D) and two loops (CD and FG) were employed for the diversified positions in a side-and-loop library [1–3]. Using this unique protein scaffold for the screening campaign by display technology, monobodies have been used in life science research, including molecular biology and

N. Iwamoto, *Synthesis of Functional Characterization of Protein Domains for the Development of Mirror-Image Therapeutics*, Springer Theses,
https://doi.org/10.1007/978-981-95-5366-2_4

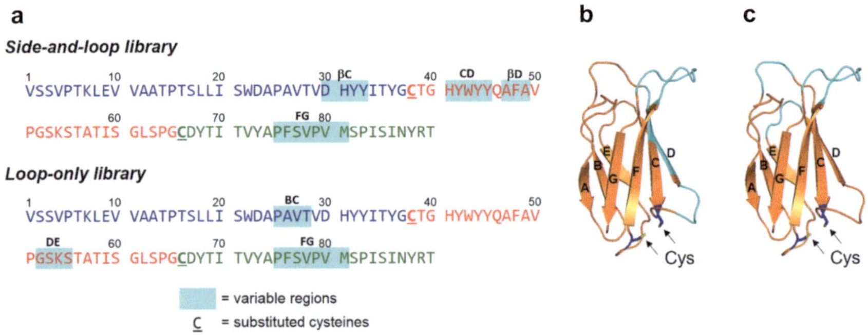

Fig. 4.1 Sequences of monobodies and design of a monobody variant with two cysteine substitutions. **a** Comparison of the variable regions in the side-and-loop and loop-only libraries. The substituted cysteines are underlined. The sequence is based on the anti-GFP monobody GS2. **b** Variable regions in the side-and-loop library in the structure of the monobody. Variable regions are colored with cyan, and the arrows indicate the substituted cysteines. **c** Variable regions in the loop-only library in the structure of the monobody

crystallography [3]. Additionally, several FN3-based therapeutic candidates have been identified for various target molecules, such as vascular endothelial growth factor receptor 2 (VEGFR2) [4] and proprotein convertase subtilisin/kexin type 9 [5]. For medicinal applications, a monobody would be advantageous over other protein scaffolds because the sequence designed from endogenous fibronectin is less likely to induce an immune response after in vivo administration [6]. However, using human-derived sequences for protein therapeutics does not necessarily eliminate the generation of anti-drug antibodies (ADAs), which may impair the therapeutic effects and/or sometimes cause adverse effects [7]. Actually, it was reported that the administration of CT-322 [8], a pegylated FN3-based protein engineered to bind VEGFR2, caused ADA generation in more than half of the patients [9].

The author has focused on developing mirror-image monobodies for novel protein therapeutics to overcome this immunogenicity shortcoming. As described in Preface, mirror-image peptides (D-peptides) and proteins (D-proteins) are expected to have favorable pharmacokinetic and safety profiles. In this study, the author investigated the potential of the mirror-image monobody as a suitable less immunogenic non-antibody scaffold in drug discovery [10]. Facile synthetic process of monobody variants was established for comparative structural and biological analyses between native and mirror-image monobodies. The designed variant was also applied to a structure–activity relationship study to identify a novel monobody scaffold with favorable molecular properties.

Chemical Synthesis of an Anti-GFP Monobody Variant. As an initial case study, the author selected a monobody, GS2, that binds with green fluorescent protein (GFP) [2]. This anti-GFP monobody, GS2, was identified by phage-display selection from a side-and-loop library. To realize efficient construction of desired monobody sequences identified by mirror-image phage display screening, the author designed a novel monobody variant, in which two cysteine substitutions were introduced into the sequence to facilitate the synthetic process by stepwise native chemical ligations (NCLs) [11]. The author chose Glu38 in the CD-loop and Val66 in the EF-loop of GS2 for substitution with cysteines, which are located in close proximity to potentially form a disulfide bond (Fig. 4.1). A previous study on Centyrin, an FN3-based protein scaffold, revealed that single cysteine mutations at the corresponding residues had relatively small effects on the target binding [12]. These two residues are also located outside of the variable regions in a side-and-loop library of monobodies and on the opposite side of the variable regions (BC-, DE-, and FG-loops) in a loop-only library of monobodies. Therefore, the overall synthetic process can be designed without dependence on the hit bioactive sequences in the variable regions. Additionally, because the upstream residues of Glu38 and Val66 are glycines (Gly37 and Gly65) in the sequence of GS2, substitutions with cysteines enable efficient NCLs at the Gly–Cys junction [13]. Thus, this modified monobody is a non-antibody scaffold that combines sufficient target binding and less synthetic effort, which is suitable for a mirror-image screening strategy.

The author investigated two synthetic routes for modified GS2 (mGS2) with cysteine substitutions. The author's initial attempt to synthesize mGS2 via an N-to-C NCL strategy is depicted in Fig. 4.2a. The N-terminal peptide segment L-**1** and middle segment L-**2a** were constructed by Fmoc-based solid-phase peptide synthesis on *N*-acyl-*N'*-methyl-benzimidazolinone (MeNbz) [14] and diaminobenzoic acid (Dbz) [15] linkers, respectively. To improve the solubility of the segments during NCL and purification, a solubilizing auxiliary sequence of four arginines was appended to the C-terminus of peptides L-**1** and L-**2a**. The C-terminal segment L-**3** was also synthesized, which contained a histidine tag at the C-terminus for detection and immobilization. 1,2,4-Triazole-mediated NCL [16] between peptides L-**1** and L-**2a** provided the intermediate L-**4**. $NaNO_2$-mediated activation of the Dbz moiety [17] in L-**4** followed by NCL in the presence of 4-mercaptophenylacetic acid (MPAA) with the C-terminal segment L-**3** afforded the expected full sequence of mGS2 (L-**6**). However, chromatographic separation of the product from a trace hydrolysate impurity from L-**4** was difficult.

Next, the author investigated the synthesis of mGS2 via a C-to-N NCL strategy (Fig. 4.2b). Because the N-terminal and C-terminal peptide segments (L-**1** and L-**3**) can be used for the C-to-N NCL strategy, only the middle segment (L-**2b**) was newly designed, which contains 1,3-thiazolidine-4-carboxylic acid (Thz) for N-terminal temporary protection and a MeNbz linker for C-terminal activation. The resulting peptide segments were assembled in the C-to-N direction. 1,2,4-Triazole-mediated NCL between peptides L-**2b** and L-**3** followed by methoxyamine-mediated deprotection of thiazolidine provided the intermediate L-**7**. Subsequently, peptide L-**7** was conjugated with the N-terminal segment L-**1** in the presence of 1,2,4-triazole again to give L-mGS2 (L-**6**) in 35% overall yield (two steps from the peptide segment L-**3**). The C-to-N NCL strategy was more straightforward than the N-to-C process because the enhanced solubility of L-**7** with a C-terminal histidine tag facilitated the purification processes after NCLs. This simple synthetic process was also used to

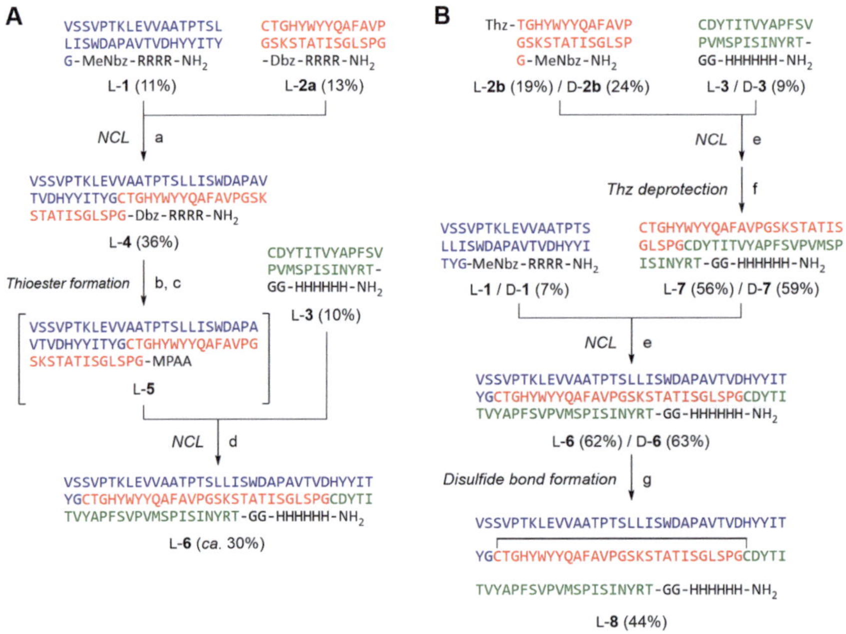

Fig. 4.2 Synthesis of anti-GFP monobody (L/D-mGS2) via N-to-C native chemical ligations (**A**) and C-to-N native chemical ligations (**B**). *Reagent and conditions*: **a** 1,2,4-triazole, TCEP, 6 M guanidine (pH 7.0); **b** $NaNO_2$, 6 M guanidine, phosphate buffer (pH 3.0); **c** MPAA, TCEP, 6 M guanidine, phosphate buffer (pH 6.2); **d** MPAA, TCEP, 6 M guanidine, phosphate buffer (pH 6.5); **e** 1,2,4-triazole, TCEP, 6 M guanidine (pH 7.0); **f** methoxyamine; **g** 2-PDS, 6 M guanidine (pH 8.0)

synthesize the mirror-image D-mGS2 (D-**6**) using D-amino acids and glycine (37% overall yield).

Of note, the author's initial attempt to establish the synthetic protocol of the monobody scaffold was conducted for the native sequence of GS2 (Fig. 4.3). Because the monobody sequence does not contain conserved Cys residues, NCLs and a desulfurization strategy were employed for ligations at the Xaa–Ala sites [18]. The sequence of GS2 was successfully constructed from four peptide segments; however, temporary modification with a solubilizing auxiliary Trt-K_{10} [19] was also required to improve the low solubility of the intermediate and purification by HPLC after NCLs. The author postulated that these steps would hamper the efficient construction of monobody sequences, and thus was motivated to investigate other synthetic approach mentioned above. The author's success in the efficient preparation of full-length mGS2 supports the validity of the design and the synthetic process for monobody variants.

Structural and Biological Evaluations of Anti-GFP Monobody Variants. With the full length of L-mGS2 (L-**6**) in hand, the author investigated the folding conditions to obtain bioactive L-mGS2. There is a reported procedure for purification of an FN3-derived binding protein from *E. coli* inclusion bodies via refolding by dialysis in

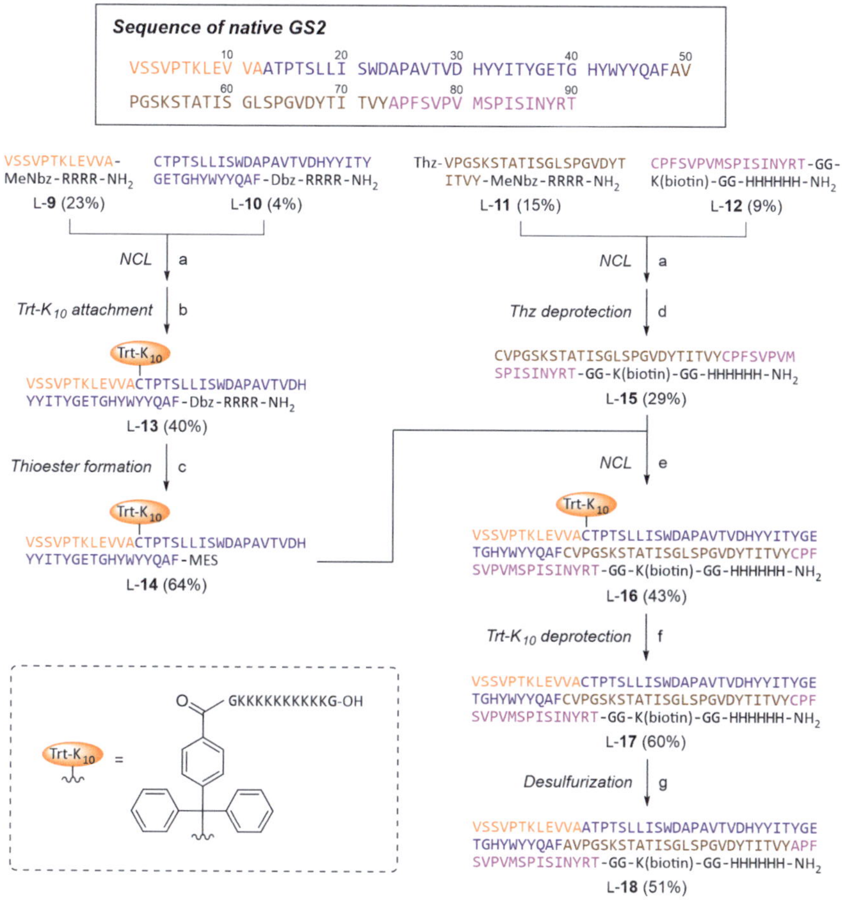

Fig. 4.3 Synthesis of biotinylated native anti-GFP monobody (L-GS2^{bitoin}). *Reagent and conditions*: **a** 1,2,4-triazole, TCEP, 6 M guanidine (pH 7.0); **b** Trt(OH)-K_{10}, TFA; **c** $NaNO_2$, 6 M guanidine, phosphate buffer (pH 3.0); and then MPAA, TCEP (pH 6.0); **d** methoxyamine; **e** MPAA, TCEP, 6 M guanidine, phosphate buffer (pH 6.6); **f** TFA/EDT (95:5); **g** VA-044, MESNa, TCEP, 6 M guanidine, phosphate buffer (pH 6.5)

acetate buffer (pH 4.5) [8]. According to the reported protocol [8] with some modifications, L-mGS2 was subjected to dialysis procedures under slightly acidic conditions (50 mM acetate buffer, pH 4.5, 5 mM TCEP) to provide the folded protein with substituted cysteines in a reduced form. Of note, when the author initially investigated various folding conditions and procedures under neutral pH, L-mGS2 was highly prone to aggregation, and obtaining bioactive L-mGS2 failed. It was reported that FN3 modules and FN3-derived proteins are thermodynamically more stable under acidic conditions compared with neutral conditions [1, 20–22]. In the current case, the slightly acidic conditions (pH 4.5) would contribute to the formation of stable structure and thus the rapid folding kinetics [20], which presumably prevented the

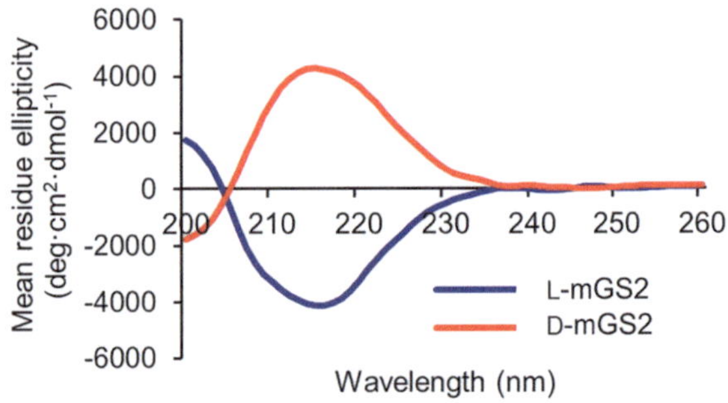

Fig. 4.4 CD spectra of L-mGS2 and D-mGS2

unfavorable self-association. The CD spectrum of folded L-mGS2 suggested the existence of a β-sheet structure, which is a common feature among FN3-derived binding proteins (Fig. 4.4). D-mGS2, which was subjected to identical folding conditions, showed an inverted CD spectrum to L-mGS2. These results suggested that synthetic D-mGS2 folded correctly to form the mirror-image structure of L-mGS2.

The bioactivity of synthetic mGS2 proteins was evaluated by surface plasmon resonance (SPR) analysis. Biotin-labeled mGS2 (L-mGS2^{biotin} and D-mGS2^{biotin}) were designed and synthesized for immobilization on the sensor chip (Fig. 4.5). Enhanced green fluorescent protein (EGFP) at various concentrations was flowed over the sensor chip. Synthetic L-mGS2^{biotin} showed slightly less potent binding toward EGFP ($K_D = 9.8 \pm 4.1$ nM) compared with biotin-labeled native GS2 without Cys substitution (L-GS2^{biotin}; $K_D = 1.8 \pm 1.3$ nM) (Table 4.1). The binding affinity of these synthetic proteins was comparable to that of recombinant GS2, which was reported previously [2]. In contrast, D-mGS2^{biotin} did not bind EGFP, suggesting that the molecular recognition of synthetic L-mGS2 with EGFP was accomplished stereoselectively.

Next, the immunogenic properties of folded L-mGS2 and D-mGS2 were investigated. BALB/c mice were immunized intraperitoneally with synthetic L- or D-mGS2 three times in combination with a Freund's adjuvant at days 0, 14 and 28. Plasma samples were collected from mice at days 0, 14, 28, 35 and 44, and the generation of an anti-L-mGS2 antibody or anti-D-mGS2 antibody [anti-drug antibody (ADA) against mGS2] was measured by ELISA. Repeated injections caused a gradual increase in the ADA level for L-mGS2-immunized mice, whereas no increase in ADA was observed for D-mGS2-immunized mice (Fig. 4.6a). The production of ADA was observed in all L-mGS2-immunized mice on day 44 after the initial administration of the monobody, with some variation in the production levels (Fig. 4.6b). In contrast, no ADA generation was observed in all D-mGS2-immunized mice. These results suggest that the mirror-image monobodies are promising protein scaffolds with less immunogenic properties and support previous reports showing that mirror-image peptides and proteins have lower immunogenic properties [23–25].

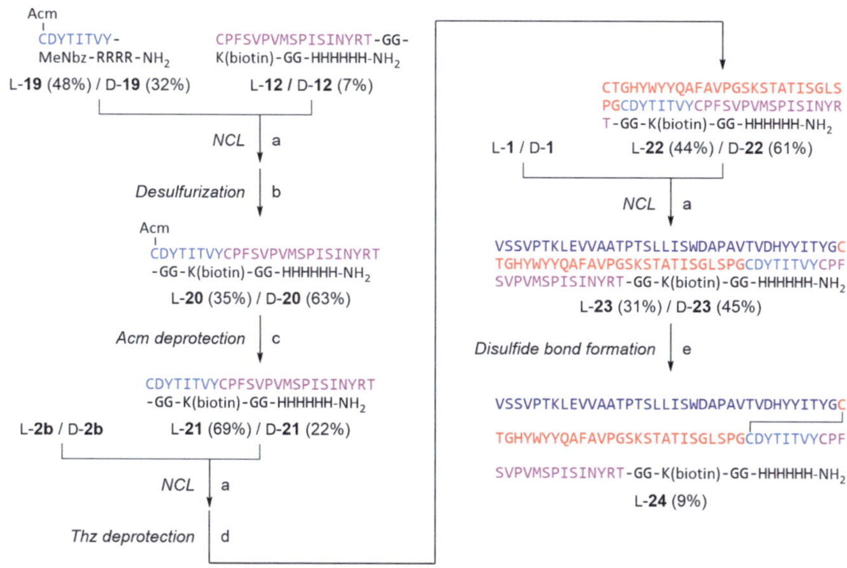

Fig. 4.5 Synthesis of biotinylated anti-GFP monobody (L/D-mGS2^{biotin} and L-mGS2$^{SS/biotin}$). *Reagent and conditions*: **a** 1,2,4-triazole, TCEP, 6 M guanidine (pH 7.0); **b** VA-044, MESNa, TCEP, 6 M guanidine, phosphate buffer (pH 6.5); **c** AgOTf, anisole, TFA; and then 10 mM DTT, 6 M guanidine, phosphate buffer (pH 7.4); **d** methoxyamine; **e** 2-PDS, 6 M guanidine, Tris buffer (pH 8.0)

Table 4.1 SPR analysis of synthetic monobodies binding with EGFP

Ligand	K_D (nM)[a]
L-GS2$^{biotin,\ b}$	1.8 ± 1.3
L-mGS2^{biotin}	9.8 ± 4.1
L-mGS2$^{SS/biotin}$	3.7 ± 0.3
D-mGS2^{biotin}	No binding

[a] K_D values were determined from triplicate assays. [b] K_D value of recombinant GS2 was reported to be 3.4 ± 0.2 nM [2]

Stabilization of Monobody Variant Scaffold with an Intramolecular Disulfide Bond. Disulfide bonds play an important role in the folding of peptide-based and protein-based therapeutics and possibly contribute to improving pharmacological properties [26]. Gilbreth et al. previously introduced a new disulfide bond between two mutated cysteines of the third fibronectin type III domain of human tenascin-C to improve the thermodynamic stability and resistance to thermolysin-mediated proteolysis [27]. In the author's newly designed monobody scaffold, because the substituted cysteines, Cys38 and Cys66, are adjacent to each other in the folded state, the author expected that an intramolecular disulfide bond between these vicinal cysteines would similarly improve protein stability. Thus, the author investigated the formation of an intramolecular disulfide bond in mGS2 and subsequent refolding for preparing the

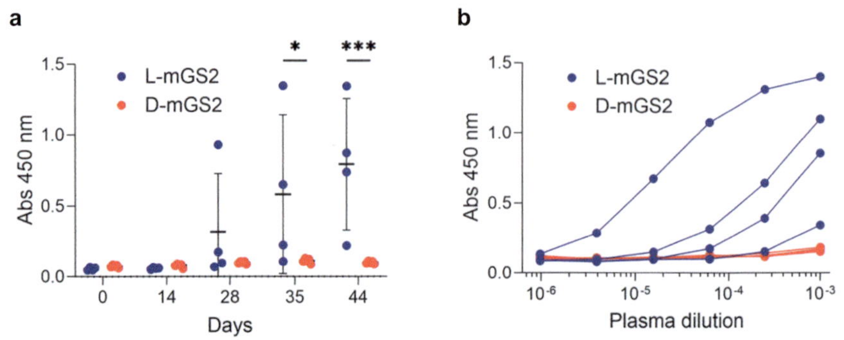

Fig. 4.6 Evaluation of the immunogenicity of L-mGS2 and D-mGS2. **a** Generation of an anti-drug antibody (ADA) in mouse sera at days 0, 14, 28, 35 and 44 after injection of L-mGS2 and D-mGS2 was detected by ELISA (L-mGS2: $n = 4$; D-mGS2: $n = 5$). Absorbance of 3,3′,5,5′-tetramethylbenzidine (TMB) was measured at 450 nm. Statistical analysis was performed by two-way ANOVA followed by Sidak's multiple comparisons test. * $p < 0.05$, *** $p < 0.001$. **b** ADA production in mouse plasma at day 44 after injection of L-mGS2 and D-mGS2. ADA in serially diluted plasma was detected by ELISA. Each line connects the data of individual mice

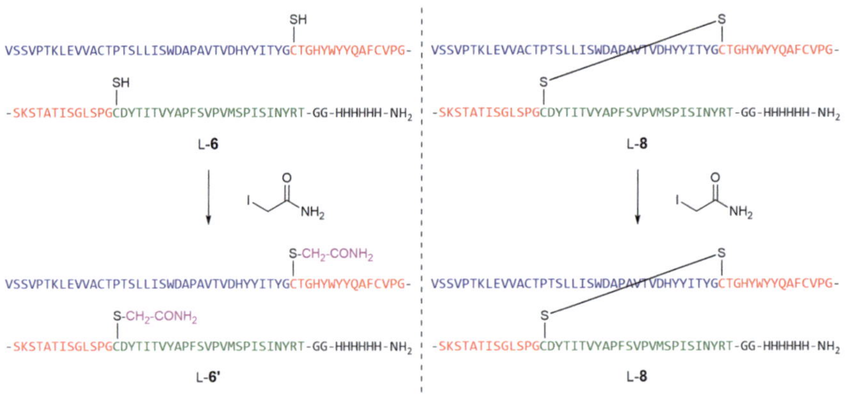

Fig. 4.7 Confirmation of disulfide bond formation in L-mGS2 and L-mGS2^{SS}. After treating peptides L-mGS2 (L-**6**) or L-mGS2^{SS} (L-**8**) with iodoacetamide, the products were analyzed by LC–MS

Peptide	Formula	Calcd	Observed m/z
L-**6**	$C_{489}H_{718}N_{124}O_{139}S_3$	10,647.21	$[M + 7H]^{7+} = 1522.50$ $[M + 8H]^{8+} = 1332.29$
L-**6'**	$C_{493}H_{724}N_{126}O_{141}S_3$	10,761.25	$[M + 7H]^{7+} = 1538.63$ $[M + 8H]^{8+} = 1346.55$
L-**8**	$C_{489}H_{716}N_{124}O_{139}S_3$	10,645.19	$[M + 7H]^{7+} = 1522.31$ $[M + 8H]^{8+} = 1332.11$
L-**8** after treatment	$C_{489}H_{716}N_{124}O_{139}S_3$	10,645.19	$[M + 7H]^{7+} = 1522.67$ $[M + 8H]^{8+} = 1331.63$

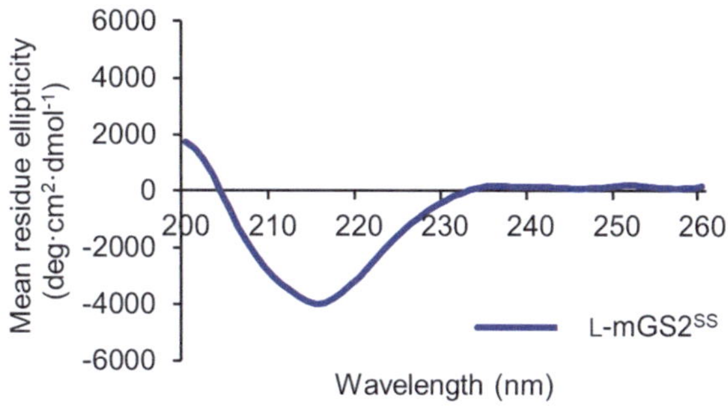

Fig. 4.8 CD spectra of L-mGS2^{SS}

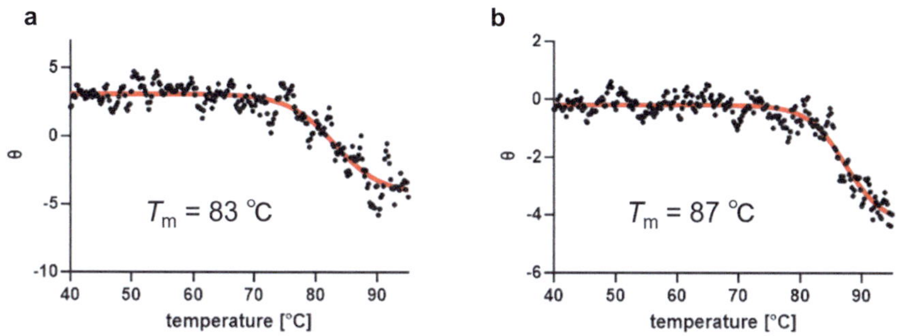

Fig. 4.9 Thermal stability of **a** L-mGS2 and **b** L-mGS2^{SS}. The thermal unfolding transition was monitored by the CD signal at 203 nm in acetate buffer. Data were smoothed with centered moving averages. The T_m of the unfolding curves was determined by data fitting to the Boltzmann sigmoidal curve using GraphPad Prism10

bioactive monobody. L-mGS2 (L-**6**) was treated with 2,2'-dithiodipyridine (2-PDS) [28] under denaturing conditions (pH 8.0) in guanidine buffer to give L-mGS2^{SS} (L-**8**) with an intramolecular disulfide bond in 44% yield (Fig. 4.2b). The formation of a disulfide bond in L-**8** was confirmed by MS measurement after treatment with iodoacetamide (Fig. 4.7). Peptide L-**8** was subjected to folding conditions in a mildly acidic buffer (50 mM acetate buffer, pH 4.5) to provide folded L-mGS2^{SS}. The β-sheet structure was verified by CD spectroscopy, which was identical to L-mGS2 (Fig. 4.8). Additionally, the thermal stability of L-mGS2 and L-mGS2^{SS} was evaluated by monitoring changes to the CD signal at 203 nm in acetate buffer (Fig. 4.9). L-mGS2^{SS} showed slightly higher thermal stability than L-mGS2. In the SPR analysis, biotin-labeled L-mGS2^{SS} (L-mGS2$^{SS/biotin}$) exhibited improved binding affinity with EGFP ($K_D = 3.7 \pm 0.3$ nM) compared with L-mGS2^{biotin} and comparable to L-GS2^{biotin} (Table 4.1). This observation suggests that the formation of the disulfide bond stabilized the bioactive form of the anti-GFP monobody and/or restored the unfavorable effect(s) of two sulfhydryl groups in L-mGS2 on binding with EGFP.

Design and Synthesis of Anti-MCP-1 Monobody Variants. A designed monobody variant with cysteine substitutions was applied to another target-binding sequence.

The D-enantiomer of Mb5-11 (D-Mb5-11) is a mirror-image monobody variant, which binds to the native monocyte chemoattractant protein 1 (MCP-1) to inhibit the interaction of MCP-1 and the C–C chemokine receptor type 2 (CCR2). Because the MCP-1/CCR2 axis plays an important role in the immune response against infectious diseases and malignant tumors [29, 30], inhibitory molecules would be potential therapeutic agents to prevent the progression of these diseases. The author next investigated the structure–activity relationship of cysteine substitutions on the structure and stability of the monobody scaffold of Mb5-11 (Fig. 4.10a) [31], which was identified by mirror-image TRAP display [32] screening against D-MCP-1 from a loop-only library of monobodies with variable regions on the BC- and FG-loops. The author designed three Mb5-11 variants with different pairs of cysteine substitutions, namely mMb5-11A, B, and C (Fig. 4.10b, c). mMb5-11A included substitutions at Arg40 in the β-strand C and Tyr80 in the neighboring β-strand F. In mMb5-11B, Thr42 and Thr78 in the β-strands C and F, respectively, were replaced with cysteines. In mMb5-11C, Glu45 in the CD-loop and Val73 in the FG-loop were substituted with cysteines, in which a similar pattern of modifications with mGS2 was introduced. In all the variants, the cysteines were located structurally vicinal to each other for potential disulfide bond formation, as well as outside of the variable regions in both the loop-only and side-and-loop monobody libraries.

Taking advantage of the synthetic process for mGS2, identical C-to-N sequential NCLs were designed for the synthesis of the 109-residue Mb5-11 variant L-**29a** from three peptide segments (Fig. 4.11a). In brief, the N-terminal and middle peptide segments (L-**25a** and L-**26a**) were constructed by Fmoc-SPPS on hydrazide-linked resin [33, 34]. The resulting crude peptide hydrazides were directly applied to acetylacetone (acac)-mediated hydrazide activation [35] to isolate the MPAA thioesters. The N-terminal cysteines of the middle peptide segment were temporarily protected as Thz groups. The C-terminal peptide segment L-**27a** was synthesized with polyhistidine tags at the C-terminus. The resulting peptide segments were then ligated in the C-to-N direction. Peptides L-**26a** and L-**27a** were subjected to MPAA-mediated NCLs followed by one-pot Thz deprotection to provide peptide L-**28a**. Further NCLs between peptides L-**25a** and L-**28a** in MPAA ligation buffer provided the full-length sequence of the Mb5-11 variant L-**29a**. The other Mb5-11 variants L-**29b** and L-**29c** were synthesized via a similar process (Fig. 4.11b, c). Notably, NCLs at sterically-hindered sites, such as Val–Cys and Ile–Cys junctions [13], were performed with an extended reaction time in MPAA ligation buffer with increased TCEP concentrations [100 mM MPAA, 100 mM TCEP, 6 M guanidine, 200 mM phosphate (pH 7)]. All the NCLs proceeded well to provide the ligated products in sufficient yields.

For an experimental control, native Mb5-11 was also synthesized (Fig. 4.11d). From the experience of the difficult synthesis of native GS2 using standard NCL and desulfurization methods (Fig. 4.3), the synthesis of native Mb5-11 was conducted by the thioester coupling method [36]. The sequence was divided into three peptide segments at Gly–Xaa sites so that no epimerization at the C-terminal amino acids occurred during fragment coupling. The peptide segments were assembled in the C-to-N direction by coupling between peptide thioesters and N-terminal primary amines using 3,4-dihydro-3-hydroxy-4-oxo-1,2,3-benzotriazine (HOOBt)/DIEA in

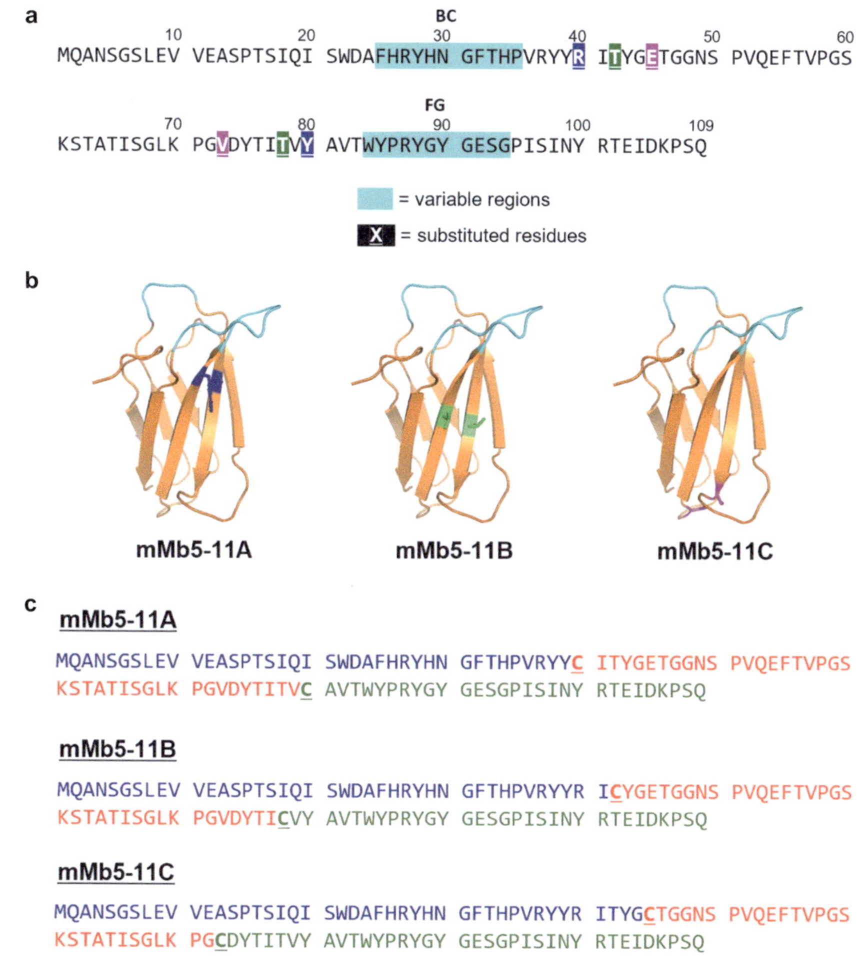

Fig. 4.10 Design of monobody variants with two cysteine substitutions. **a** Sequence of Mb5-11. The variable regions are highlighted in cyan. The residues for cysteine substitutions are highlighted with different colors (blue: mMb5-11A, green: mMb5-11B, pink: mMb5-11C). **b** Schematic view of the designed monobody variants. The substituted residues in the variants are shown in different colors. **c** Sequences of Mb5-11 variants

DMSO. A hydrophilic iNoc group [37] was employed for protection of the primary amino groups at the N-terminus of peptide L-**30** and the lysine side chains in the other peptide segments (L-**31** and L-**32**). Using this iNoc protection, site-selective couplings at the Gly44–Val45 and Gly91–Glu92 sites were possible [36]. Native Mb5-11 (L-**35**) was successfully obtained after a two-step coupling followed by deprotection of the iNoc groups under reducing conditions, although the preparation of monobodies with a variety of sequences was hampered by the use of the non-standard iNoc and hydrophobic Fmoc protecting groups.

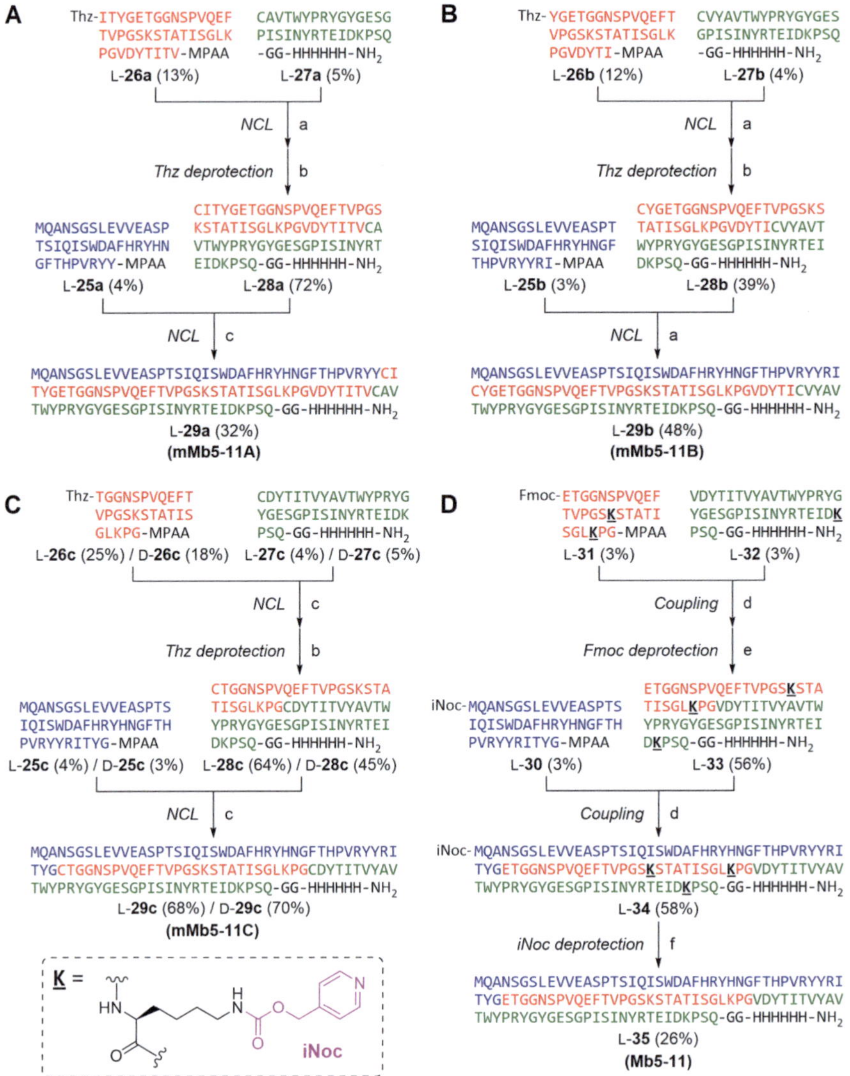

Fig. 4.11 Synthesis of **A** mMb5-11A, **B** mMb5-11B, **C** mMb5-11C, and **D** Mb5-11. **K** indicates the iNoc-protected lysine. *Reagent and conditions*: **a** 100 mM MPAA, 50 mM TCEP, 6 M guanidine, 200 mM phosphate (pH 7); **b** methoxyamine; **c** 100 mM MPAA, 100 mM TCEP, 6 M guanidine, 200 mM phosphate (pH 7); **d** HOOBt, DIEA, DMSO; **e** piperidine; **f** Zn dust, AcOH/H_2O (1:1)

Effect of the Substitution Positions on the Bioactivity and Thermal Stability of Monobody Variants. With synthetic Mb5-11 and the variants (mMb5-11A–C) in hand, the target binding was comparatively evaluated (Fig. 4.12a). The Mb5-11 variants were subjected to dialysis in acetate buffer (pH 4.5) according to the refolding procedure of mGS2 to provide the folded proteins. The binding of monobody variants to D-MCP-1 was evaluated by SPR analysis under neutral conditions (pH 7.4), in which biotinylated D-MCP-1 was immobilized on a streptavidin (SA) sensor chip. mMb5-11A and mMb5-11B having Cys substitutions on two β-strands exhibited slightly lower binding affinity toward D-MCP-1 compared with native Mb5-11 [K_D(Mb5-11) = 1.3 ± 0.2 nM; K_D(mMb5-11A) = 2.9 ± 0.9 nM; K_D(mMb5-11B) = 4.8 ± 0.2 nM]. In contrast, mMb5-11C with Cys substitutions at two loops possessed equipotent affinity with the native form [K_D(mMb5-11C) = 1.2 ± 0.1 nM].

The thermal stability of mMb5-11A–C was analyzed by monitoring changes in the CD signals at 203 nm with increasing temperature in phosphate buffer (pH 7.0) (Fig. 4.12b). The thermal stability of mMb5-11A and mMb5-11C was slightly lower than that of native Mb5-11 [T_m(Mb5-11) = 79 °C; T_m(mMb5-11A) = 71 °C; T_m(mMb5-11C) = 72 °C], while mMb5-11B showed similar thermal stability to native Mb5-11 [T_m(mMb5-11B) = 81 °C]. Thus, the binding affinity of the anti-D-MCP-1 monobody variants was not correlated with the thermal stability. These observations of anti-D-MCP-1 monobody variants demonstrated that the molecular properties could be adjusted by selecting the Cys-substitution positions.

Subsequently, the author investigated the intramolecular disulfide bond formation of the two introduced cysteine residues in the Mb5-11 variants. The folded proteins (SH forms) were subjected to oxidative conditions with cystamine at neutral pH, and the resulting samples were analyzed by LC–MS (Fig. 4.13). Formation of intramolecular disulfide bonds from mMb5-11C with Cys-substitutions at two loop regions was observed, while the reactions of mMb5-11A and mMb5-11B provided cysteamine adducts via intermolecular disulfide bond formation. This result would presumably be attributable to disulfide exchange of the monobody variants with cystamine. Previously, it has been reported that disulfide bond formation between a pair of hydrogen-bonded cysteines across antiparallel β-strands was less favorable, compared with a non-hydrogen-bonded pair of cysteines across β-strands, which was more likely to result in the formation of a disulfide bond to stabilize the protein structure [38]. In the monobodies, mMb5-11A and mMB5-11B, the two cysteines would face each other via hydrogen bonding across two β-strands (strands C and F), which would prevent the expected formation of intramolecular disulfide bonds even under the oxidative conditions.

Modification of Cysteine Sulfhydryl Groups in Monobody Variants. On the basis of the different profiles of disulfide bond formation among the Mb5-11 variants, the author devised an alternative strategy to modify the cysteine residues to stabilize the monobody scaffolds as well as improve the molecular properties. A number of cysteine-selective modification strategies have been developed for bioconjugation using the reactivity of the cysteine sulfhydryl group [39]. As a representative example, maleimide has been employed for the cysteine-selective modification of protein therapeutics with a wide variety of functional moieties [40]. Certolizumab pegol is an anti-TNFα Fab' fragment modified with a 40 kDa polyethylene glycol (PEG) moiety via a cysteine–maleimide linkage [41]. The application of these cysteine-selective

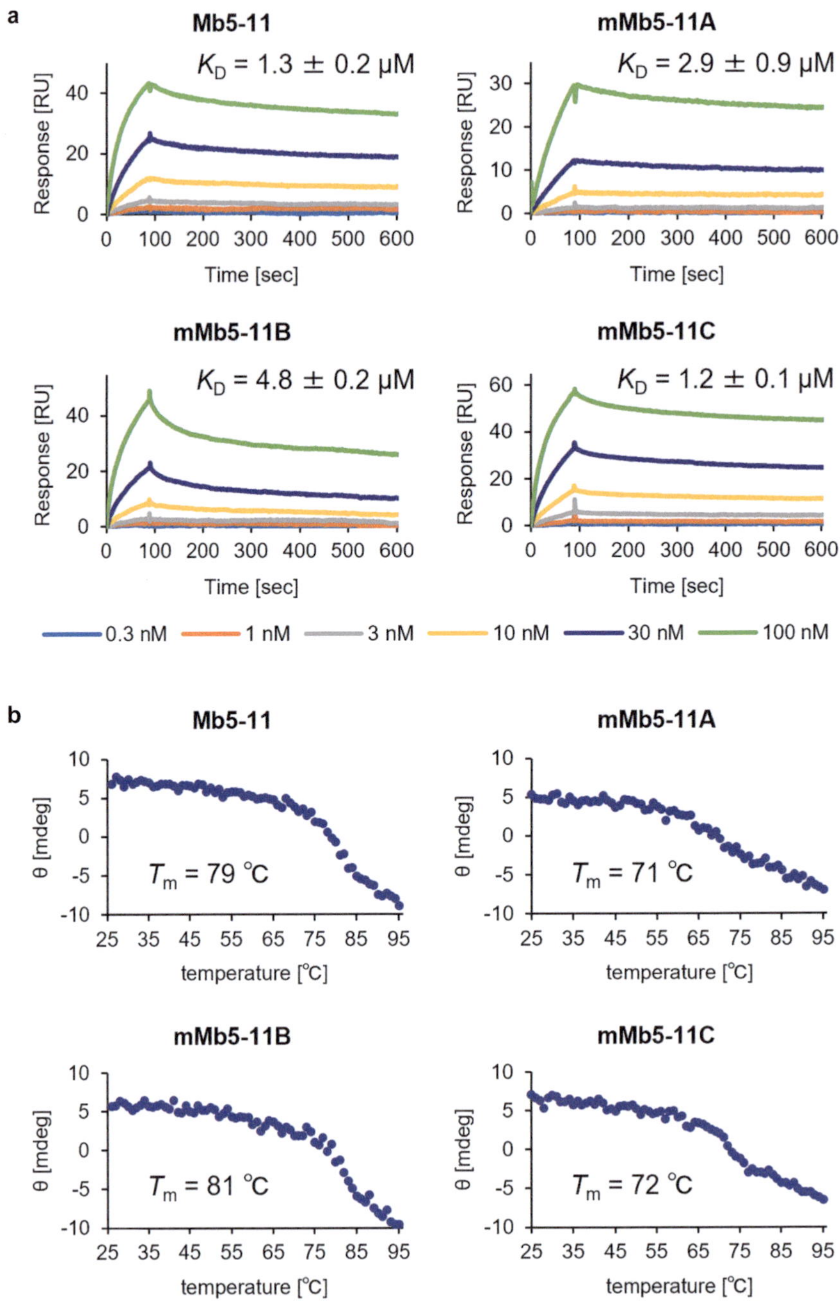

Fig. 4.12 Effects of cysteine substitutions in the scaffolds of Mb5-11 variants. **a** Representative sensorgrams of SPR analysis of synthetic monobody variants for D-MCP-1. K_D values were determined from triplicate assays. **b** Transitions of CD signals of synthetic monobody variants at 203 nm. T_m values were determined as the temperatures at the time points when half of the protein molecules were unfolded

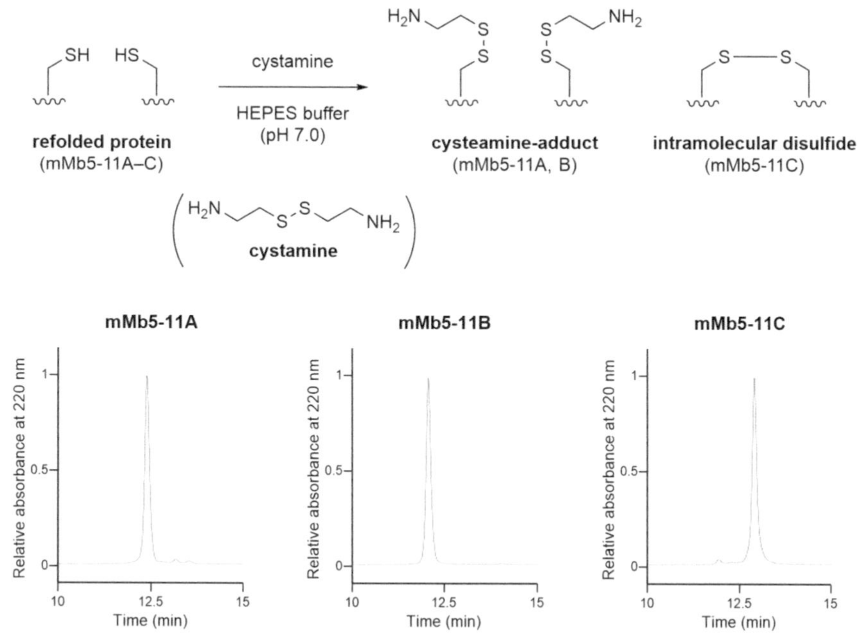

Fig. 4.13 Investigation of disulfide bond formation between two cysteine residues in folded Mb5-11 variants. After oxidation using cystamine, the products were analyzed by LC–MS. Calculated masses and observed *m/z* values for each product are shown in the table. *HPLC conditions*: Cosmosil 5C18-AR300 column with a linear gradient of 25–45% CH_3CN containing 0.05% TFA at a flow rate of 1 mL/min over 20 min

Peptide	Formula	Calcd	Observed *m/z*
mMb5-11A	$C_{580}H_{859}N_{163}O_{175}S_5$	13067.19	$[M + 8H]^{8+} = 1634.50$ $[M + 9H]^{9+} = 1453.05$
mMb5-11B	$C_{587}H_{866}N_{166}O_{174}S_5$	13184.26	$[M + 8H]^{8+} = 1648.95$ $[M + 9H]^{9+} = 1465.95$
mMb5-11C	$C_{581}H_{852}N_{164}O_{174}S_3$	13006.20	$[M + 8H]^{8+} = 1626.80$ $[M + 9H]^{9+} = 1446.15$

methods to monobody variants is expected to offer new opportunities to append a variety of additional functions to proteins.

Considering the different profiles of disulfide bond formation among the folded Mb5-11 variants, denaturing conditions using guanidine solution were employed for cysteine derivatization using modifying reagents to eliminate the effect of the possible different reactivity of the folded proteins (Fig. 4.14). Initially, the cysteines in L-**29a–c** were converted into alanine by desulfurization [18] to provide Ala-substituted derivatives L-**36a–c**. Treatment with iodoacetamide provided acetamide derivatives

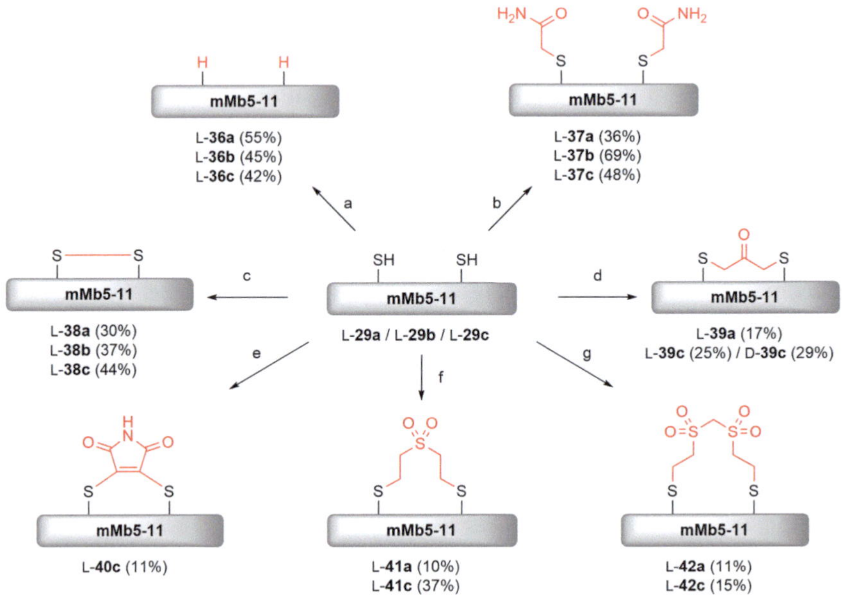

Fig. 4.14 Functionalization of cysteine sulfhydryl groups in monobody variants. *Reagent and conditions*: **a** VA-044, MESNa, TCEP, 6 M guanidine, phosphate buffer (pH 6.5); **b** iodoacetamide, 6 M guanidine, 200 mM phosphate (pH 7); **c** DTNB, 6 M guanidine, Tris buffer (pH 8); **d** 1,3-dichloroacetone, 6 M guanidine, Tris buffer (pH 8); **e** divinylsulfone, 6 M guanidine, Tris buffer (pH 8); **f** bis(vinylsulfonyl)methane, 6 M guanidine, Tris buffer (pH 8); **g** 3,4-dibromomaleimide, 6 M guanidine, Tris buffer (pH 8)

L-**37a–c**. Disulfide derivatives L-**38a–c** were synthesized using 5,5'-dithiobis(2-nitrobenzoic acid) (DTNB) [42] under denaturing conditions. Unlike in the case of the cysteamine-mediated disulfide bond formation of folded proteins, the efficient formation of intramolecular disulfide bonds was observed in all the variants L-**29a–c**. After purification by preparative HPLC, the isolated products were subjected to folding conditions to afford functional proteins for further analyses.

The binding affinities and thermal stabilities of folded Mb5-11 variants L-**36–38** were measured by SPR analysis and temperature-dependent CD analysis, respectively (Table 4.2). Ala-substituted derivatives L-**36a–c** had nearly equipotent binding to D-MCP-1 compared with the parent Mb5-11 variants L-**29a–c** [K_D(L-**36a**) = 2.1 ± 0.2 nM; K_D(L-**36b**) = 3.6 ± 0.1 nM; K_D(L-**36c**) = 1.1 ± 0.1 nM]. Modification of mMb5-11A with carbamoylmethyl groups (L-**37a**) led to a decrease in the target binding with maintenance of thermal stability [K_D(L-**37a**) = 5.9 ± 0.7 nM; T_m(L-**37a**) = 72 °C]. Introduction of the same carbamoylmethyl groups onto the SH groups in mMb5-11B and mMb5-11C had no appreciable effect on the bioactivity and stability [K_D(L-**37b**) = 3.4 ± 0.3 nM; T_m(L-**37b**) = 76 °C; K_D(L-**37c**) = 1.1 ± 0.1 nM; T_m(L-**37c**) = 73 °C]. Interestingly, the disulfide derivatives showed considerably different properties depending on the position of the disulfide bond. mMb5-11C

Table 4.2 Modification of two cysteine residues in monobody variants

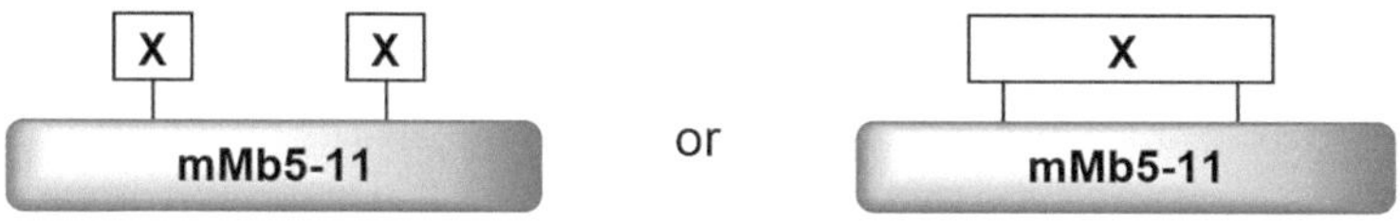

Protein	Scaffold	Substitutions	X	K_D (nM)[a]	T_m (°C)[b]
L-**35**	Mb5-11			1.3 ± 0.2	79
L-**29a**	mMb5-11A	R40C/Y80C	–SH (cysteine)	2.9 ± 0.9	71
L-**36a**			–H (alanine)	2.1 ± 0.2	69
L-**37a**			$–CH_2CONH_2$	5.9 ± 0.7	72
L-**38a**			–S–S-(disulfide)	8.3 ± 0.3	40[c]
L-**29b**	mMb5-11B	T42C/T78C	SH (cysteine)	4.8 ± 0.2	81
L-**36b**			–H (alanine)	3.6 ± 0.1	79
L-**37b**			$–CH_2CONH_2$	3.4 ± 0.3	76
L-**38b**			–S–S-(disulfide)	11.3 ± 0.1	61
L-**29c**	mMb5-11C	E45C/V73C	–SH (cysteine)	1.2 ± 0.1	72
L-**36c**			–H (alanine)	1.1 ± 0.1	71
L-**37c**			$–CH_2CONH_2$	1.1 ± 0.1	73
L-**38c**			–S–S-(disulfide)	1.9 ± 0.1	72

[a] K_D values were determined by SPR analyses from triplicate assays. [b] T_m values were determined from the transitions of CD signals at 203 nm. [c] Sample containing a small amount of aggregate was used for measurement because of the low solubility in phosphate buffer

(L-**38c**) with a disulfide bond between two loop regions, showed nearly equipotent bioactivity and thermal stability to the parent mMb5-11C (L-**29c**) [K_D(L-**38c**) = 1.9 ± 0.1 nM; T_m(L-**38c**) = 72 °C]. In contrast, mMb5-11A (L-**38a**) and mMb5-11B (L-**38b**), with disulfide bonds between β-strands, showed considerable decreases in the target binding and thermal stability [K_D(L-**38a**) = 8.3 ± 0.3 nM; T_m(L-**38a**) = 40 °C; K_D(L-**38b**) = 11.3 ± 0.1 nM; T_m(L-**38b**) = 61 °C]. These profiles of scaffold stability of the disulfide derivatives were well correlated with the efficiency of the disulfide bond formation of the folded monobody variants using cystamine (Fig. 4.13), suggesting that disulfide bond formation between the two β-strands was less favorable compared with that between the two loop regions.

Next, the author focused on chemical stapling by the cysteine crosslinking of monobody variants. In chemical stapling strategies, covalent tethering between two residues can restrict the conformational flexibility of the parent peptides, leading to improvements in the molecular properties, such as target binding affinity, target selectivity, and proteolytic resistance [43]. It is well documented that chemical stapling techniques are useful in cyclic small peptides [44] and α-helical peptides [45]. The author examined the applicability of chemical stapling to the modification of Mb5-11 variants consisting of β-sandwich structures. Considering the relatively low binding affinity of the mMb5-11B derivatives (L-**29b**, L-**36b–38b**), mMb5-11A and mMb5-11C were employed for the chemical stapling experiments (Fig. 4.14). Alkylations for stapling were performed under denaturing conditions in the presence of guanidine. Stapled derivatives L-**39a,c** with an acetone linker were synthesized by treatment with 1,3-dichloroacetone [46]. Sulfone-stapled derivatives L-**40a,c** and bissulfone-stapled derivatives L-**41a,c** were synthesized by treatment of mMb5-11A and mMb5-11C with vinylsulfone reagents [47]. Reaction of mMb5-11C with 3,4-dibromomaleimide provided the maleimide-stapled derivative L-**42c** [48]. Of note, the product from the maleimide-stapling of mMb5-11A was not obtained during the HPLC purification probably because of its hydrophobic property.

These stapled derivatives were comparatively evaluated to assess the effect of the newly introduced staple linkers on the D-MCP-1 binding and the thermal stability (Table 4.3). Compared with the unstapled Mb5-11 variants, similar or slightly less potent D-MCP-1 binding was observed for the stapled derivatives [K_D(L-**39a–41a**) = 2.6–4.9 nM; K_D(L-**39c–41c**) = 1.2–1.6 nM]. The Mb5-11 variant L-**42c** with a maleimide linker between the loop regions was 2.4-fold less potent compared with the parent mMb5-11C [K_D(L-**42c**) = 2.9 ± 0.1 nM]. All the stapled derivatives L-**39a–41a** and L-**39c–42c** had slightly enhanced thermal stability compared with the parent mMb5-11A and mMb5-11C (SH forms), respectively [T_m(L-**39a–41a**) = 76–79 °C; T_m(L-**39c–42c**) = 77–86 °C]. Importantly, although disulfide bond formation in mMb5-11A lowered the thermal stability, the introduction of staple linkers into mMb5-11A maintained the stable scaffold structures. These findings suggested that the chemical crosslinking of a pair of hydrogen-bonded cysteines between antiparallel β-strands is favorable to design novel monobody scaffolds with high thermal stability. Among the monobody scaffolds investigated in the present study, the mMb5-11C variant with an acetone stapling between the loop regions (L-**39c**) was the most potent derivative and had the best thermal stability.

The author then synthesized the D-protein of the optimized monobody variant, mMb5-11C, with an acetone stapling. The synthesis, which was performed using D-amino acids following the same procedures as for the L-protein, successfully provided the D-enantiomer of acetone-stapled mMb5-11C D-**39c** (Figs. 4.11c, and 4.14). The resulting folded D-monobody variant showed a nearly inverted CD spectrum to the L-isomer (Fig. 4.15), which supported the presence of a mirror-image secondary structure. SPR analysis revealed that the D-monobody bound to L-MCP-1 with subnanomolar binding affinity [K_D(D-**39c**) = 0.7 ± 0.1 nM], but did not show binding to D-MCP-1 (Table 4.3, and Fig. 4.16). The stereoselective MCP-1-binding of the D-monobody was well reproduced in the acetone-stapled mMb5-11C [31]. Additionally, the thermal stability of the D-monobody was comparable to that of the L-enantiomer [T_m(D-**39c**) = 85 °C] (Table 4.3).

Ketone moieties in macrocyclic compounds can be used as reactive handles for bioconjugation [46, 49]. The author investigated oxime ligation on the acetone linker of the D-monobody variant D-**39c**. Under aniline-catalyzed ligation conditions [50], the refolded protein of D-**39c** was treated with a model reactant, methoxyamine (100 eq), in acetate buffer (pH 4.5) to form oxime D-**43c**, which was obtained as a

Table 4.3 Chemical stapling between two cysteine residues in monobody variants

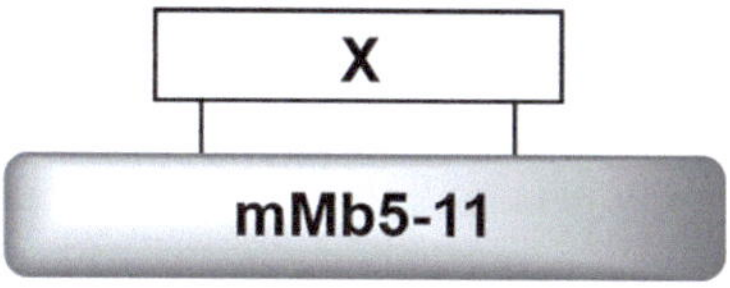

Protein		X	K_D (nM)[a]	T_m (°C)[b]
L-**35**	Mb5-11		1.3 ± 0.2	79
L-**29a**	mMb5-11A	–SH (cysteine)	2.9 ± 0.9	71
L-**39a**		$-SCH_2-CO-CH_2S-$	2.6 ± 0.2	76
L-**40a**		$-S(CH_2)_2-SO_2-(CH_2)_2S-$	3.4 ± 0.1	78
L-**41a**		$-S(CH_2)_2-SO_2-CH_2-SO_2-(CH_2)_2S-$	4.9 ± 0.4	79
L-**29c**	mMb5-11C	–SH (cysteine)	1.2 ± 0.1	72
L-**39c**		$-SCH_2-CO-CH_2S-$	1.2 ± 0.1	86
L-**40c**		$-S(CH_2)_2-SO_2-(CH_2)_2S-$	1.5 ± 0.1	82
L-**41c**		$-S(CH_2)_2-SO_2-CH_2-SO_2-(CH_2)_2S-$	1.6 ± 0.1	77
L-**42c**		–Maleimide–	2.9 ± 0.1	85
D-**39c**	D-mMb5-11C	$-SCH_2-CO-CH_2S-$	0.7 ± 0.1	85
D-**43c**		$-SCH_2-(C{=}N-OMe)-CH_2S-$	0.6 ± 0.1	83

[a] K_D values were determined by SPR analyses from triplicate assays. [b] T_m values were determined from the transitions of CD signals at 203 nm

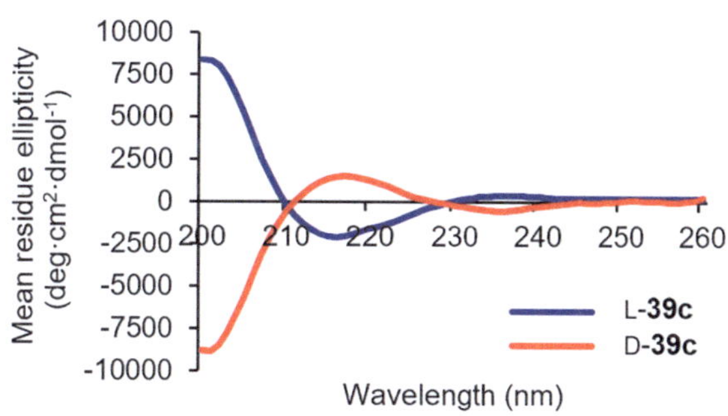

Fig. 4.15 CD spectra of L-mMb5-11C and D-mMb5-11C with acetone stapling

mixture of possible *cis* and *trans* oxime isomers (Fig. 4.17). The binding affinity to L-MCP-1 and the thermal stability of D-**43c** were comparable to those of D-**39c** [K_D(D-**43c**) = 0.6 ± 0.1 nM; T_m(D-**43c**) = 83 °C], suggesting that the oxime modification on the D-monobody variant did not impair the molecular properties. Overall, these results demonstrated that the D-monobody (D-**39c**), equipped with additional functional moieties, could be a promising therapeutic and diagnostic molecule for MCP-1-driven diseases.

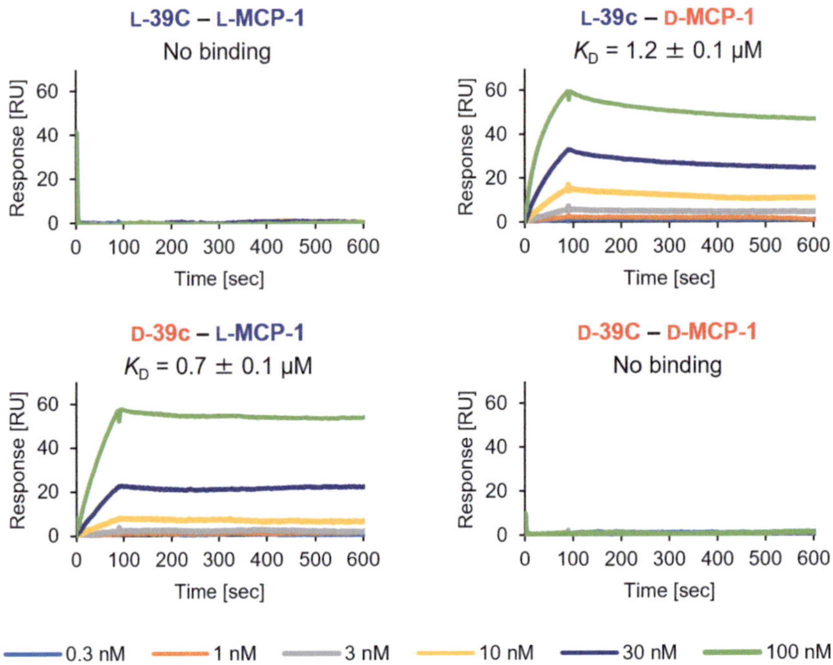

Fig. 4.16 Representative sensorgrams of SPR analysis of the synthetic monobody derivatives for L-MCP-1 and D-MCP-1. Various concentrations of L-monobody or D-monobody were injected over immobilized L-MCP-1 or D-MCP-1 (~ 100 RU). The binding affinity was determined from triplicate assays by fitting a 1:1 binding model

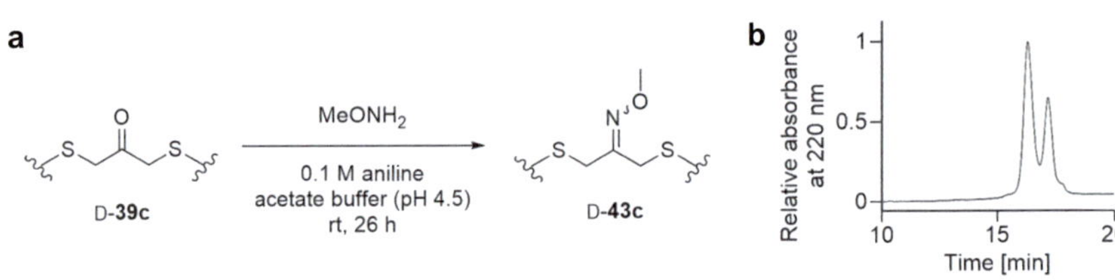

Fig. 4.17 Oxime ligation on the D-monobody variant. **a** Scheme for reaction of the acetone linker. **b** HPLC profile of the ligation products. *HPLC conditions*: Cosmosil 5C18-AR300 column with a linear gradient of 20–40% CH_3CN containing 0.05% TFA at a flow rate of 1 mL/min over 20 min

In conclusion, the author established a facile synthetic process for newly designed monobody variants with two cysteine substitutions via two-step NCLs from three peptide segments. In the synthetic process, the bioactive sequence in the side-and-loop library and loop-only library can be included in each peptide segment, providing a general synthetic approach for various monobody proteins. In the initial model study on an anti-GFP monobody GS2, the resulting synthetic L-mGS2 retained the reported structure of the recombinant protein and sufficient binding to EGFP. The immunogenicity assessment suggested that synthetic D-mGS2 showed significantly less ADA generation compared with L-mGS2. Forming an intramolecular disulfide bond in L-mGS2 was suggested to improve the binding affinity and thermal stability.

Additionally, the design of monobody variants having two cysteine substitutions was applied to an anti-D-MCP-1 monobody, Mb5-11. Three Cys-substitution patterns on the monobody scaffold were investigated in terms of target binding and scaffold stability. Taking advantage of cysteine-selective chemistries, the Mb5-11 variants were further derivatized by desulfurization, alkylation, disulfide formation, and chemical stapling. The structure–activity relationship study of the monobody scaffold revealed that the molecular properties could be altered by the modifications, resulting in identification of a potent and stable Mb5-11 variant with an acetone stapling. The synthetic D-Mb5-11 variant with an optimized modification showed potent binding to native MCP-1 as well as excellent thermal stability. These findings highlight the potential of D-monobody variants as novel therapeutic and diagnostic molecules, which could provide specific target binding, reduced synthetic effort, lower immunogenicity, and high thermodynamic stability.

4.1 Experimental Section

General

Peptide Synthesis. All reagents and solvents were purchased from Watanabe Chemical Industries, Ltd. (Hiroshima, Japan), Kokusan Chemical Industries, Ltd. (Kanagawa, Japan), Sigma-Aldrich JAPAN (Tokyo, Japan), Wako Pure Chemical Industries, Ltd. (Osaka, Japan), Tokyo Chemical Industry Co., Ltd. (Tokyo, Japan) or Nacalai Tesque, Inc. (Kyoto, Japan). For analytical HPLC, a Cosmosil 5C18-AR300 column (Nacalai Tesque, 4.6 × 250 mm) or a Cosmosil 5C4-AR300 column (Nacalai Tesque, 4.6 × 150 mm) was employed with a linear gradient of CH_3CN containing 0.1% (v/v) TFA or 0.05% (v/v) TFA at a flow rate of 1 mL/min. The products were detected by UV absorbance at 220 nm. For preparative HPLC, a Cosmosil 5C18-AR300 preparative column (Nacalai Tesque, 20 × 250 mm, flow rate 8 mL/min), a Cosmosil Protein-R preparative column (Nacalai Tesque, 20 × 150 mm, flow rate 8 mL/min) or a Cosmosil 5C4-AR300 preparative column (Nacalai Tesque, 20 × 150 mm at a flow rate 8 mL/min, or 10 × 150 mm at a flow rate 4 mL/min) was employed with a linear gradient of CH_3CN containing 0.1% (v/v) TFA or 0.05% (v/v) TFA. All peptides were characterized by ESI–MS [micromass ZQ (Waters), LCMS-2020 (Shimadzu)].

Fmoc-SPPS. Standard Fmoc-SPPS was performed using a PSSM-8 automatic peptide synthesizer (Shimadzu, Kyoto, Japan) or a CYCROSS automatic peptide synthesizer (DFC, Kyoto, Japan). The following side chain protected amino acids were employed: Arg(Pbf), Asn(Trt), Asp(O*t*-Bu), Cys(Trt), Gln(Trt), Glu(O*t*-Bu), His(Trt), Lys(Boc), Ser(*t*-Bu), Thr(*t*-Bu), Trp(Boc), Tyr(*t*-Bu). Fmoc-protected amino acids were coupled using OxymaPure (5 equiv)/DIC (5 equiv) or HBTU (5 equiv)/HOBt·H_2O (5 equiv)/DIEA (10 equiv) in DMF for 60 min twice and Fmoc-deprotection was performed by 20% piperidine in DMF for 4 min twice or 5 min twice.

Microwave-Assisted Fmoc-SPPS. Microwave-assisted peptide synthesis was performed on a Liberty BLUE (CEM Japan, Tokyo, Japan) peptide synthesizer. The following side chain protected amino acids were employed: Arg(Pbf), Asn(Trt), Asp(O*t*-Bu), Cys(Trt), Gln(Trt), Glu(O*t*-Bu), His(Trt), Lys(Boc), Ser(*t*-Bu), Thr(*t*-Bu), Trp(Boc), Tyr(*t*-Bu). Fmoc-protected amino acids were coupled using OxymaPure (5 equiv)/DIC (5 equiv) in DMF at 90 °C for 2 min, except for His(Trt). For coupling of Arg(Pbf), the reaction was repeated twice under the same conditions. His(Trt) was coupled at 50 °C for 10 min. Fmoc-deprotection was performed by 20% piperidine in DMF for 1 min at 90 °C.

Anti-GFP Monobody

Synthesis of Peptide Segments with a MeNbz Moiety at the C-terminus: Synthesis of L-mGS2$^{1–37}$ (L-1). Poly-arginine tag was manually loaded on SAL resin (0.57 mmol/g, 70 mg, 0.040 mmol) by Fmoc-Arg(Pbf)-OH (0.13 g, 0.20 mmol), OxymaPure (28 mg, 0.20 mmol) and DIC (31 μL, 0.20 mmol) in DMF for 2 h each step. MeDbz linker was then loaded on the resin by Fmoc-MeDbz-OH (78 mg, 0.20 mmol), HATU (76 mg, 0.20 mmol) and DIEA (70 μL, 0.40 mmol) in DMF for 3 h. The following Gly37 was coupled by Fmoc-Gly-OH (59 mg, 0.20 mmol), HATU (76 mg, 0.20 mmol) and DIEA (70 μL, 0.40 mmol) in DMF for 2 h. The remaining peptide sequence was constructed by a standard protocol of Fmoc-SPPS. Boc-Val-OH was employed for coupling of the N-terminal Val1. After the solid-phase peptide synthesis, a solution of *p*-nitrophenyl chloroformate in CH_2Cl_2 (90 mg/mL, 1 mL) was added to the protected peptide resin, and the reaction was continued for 2 h at room temperature. Subsequently, the resin was treated with 1 M DIEA in DMF (1 mL) for 2 h at 37 °C twice. The resin was then treated with TFA/H_2O/*m*-cresol/thioanisole/EDT (80:5:5:5:5) for 2 h at room temperature. After removal of the resin by filtration, the filtrate was poured into ice-cold dry Et_2O. The precipitate was purified by preparative HPLC (Cosmosil 5C18-AR300 column, a linear gradient of 26–56% CH_3CN in H_2O containing 0.1% (v/v) TFA over 90 min) to afford L-**1** (23.6 mg, 11% from resin). MS(ESI) *m/z*, calcd for $C_{217}H_{338}N_{60}O_{61}$: 4760.52; observed: $[M + 4H]^{4+}$ $m/z = 1191.92$, $[M + 5H]^{5+}$ $m/z = 953.76$.

L-mGS2$^{38–65}$ (L-2a). By the standard protocol for peptide synthesis, peptide L-**2a** was synthesized (9.3 mg, 13% yield) from Rink amide resin (33 mg, 0.020 mmol). MS(ESI) *m/z*, calcd for $C_{166}H_{248}N_{52}O_{44}S$: 3705.85; observed: $[M + 3H]^{3+}$ $m/z = 1237.61$, $[M + 4H]^{4+}$ $m/z = 928.38$.

[Thz^{38}]-L-mGS2$^{38–65}$ (L-2b). By the same protocol for the synthesis of peptide L-**1**, peptide L-**2b** was synthesized (23.2 mg, 19% yield) from Fmoc-NH-SAL resin (70 mg, 0.040 mmol). MS(ESI) *m/z*, calcd for $C_{145}H_{200}N_{36}O_{41}S$: 3133.44; observed: $[M + 2H]^{2+}$ $m/z = 1568.60$, $[M + 3H]^{3+}$ $m/z = 1046.15$.

L-mGS2$^{66–90}$ (L-3). Peptide L-**3** was synthesized (11.3 mg, 10% yield) from TGR resin (0.10 g, 0.025 mmol) using a microwave-assisted peptide synthesizer. MS(ESI) *m/z*, calcd for $C_{170}H_{246}N_{50}O_{45}S_2$: 3771.79; observed: $[M + 2H]^{2+}$ *m/z* = 1887.71, $[M + 3H]^{3+}$ *m/z* = 1258.67.

L-mGS2$^{1–65}$ (L-4). Peptides L-**1** (3.1 mg) and L-**2a** (2.2 mg) were dissolved in 1,2,4-triazole ligation buffer [2.5 M 1,2,4-triazole, 30 mM TCEP, 6 M guanidine (pH 7.0); 0.65 mL], and the reaction mixture was incubated for 1 h at 37 °C. After H_2O/CH_3CN (1:1) containing 0.1% (v/v) TFA was added, the mixture was purified by preparative HPLC (Cosmosil 5C18-AR300 column, a linear gradient of 27–57% CH_3CN in H_2O containing 0.1% (v/v) TFA over 90 min) to provide L-**4** (1.7 mg, 36% yield). MS(ESI) *m/z*, calcd for $C_{350}H_{529}N_{93}O_{99}S$: 7650.89; observed: $[M + 4H]^{4+}$ *m/z* = 1915.16, $[M + 5H]^{5+}$ *m/z* = 1532.32.

L-mGS2 (L-6, *N-to-C synthesis*). A $NaNO_2$ solution (200 mM, 4.3 μL) was added to a solution of L-**4** (0.38 mg) in activation buffer [6 M guanidine, 100 mM phosphate buffer (pH 3.0); 43 μL] at −20 °C, and the reaction was continued for 30 min. An MPAA solution [200 mM MPAA, 100 mM TCEP, 6 M guanidine, 200 mM phosphate buffer (pH 6.5); 15 μL] was then added to the reaction, and the reaction mixture was stirred for 5 min at −20 °C. After L-**3** (1.6 mg) in MPAA solution (28 μL, pH 7.9) was added, and the reaction mixture (pH 6.5) was incubated for 1 h at 37 °C. The mixture was purified by preparative HPLC (Cosmosil 5C4-AR300 preparative column, a linear gradient of 28–58% CH_3CN in H_2O containing 0.05% (v/v) TFA over 90 min) to provide L-**6** (0.16 mg, *ca.* 30% yield). MS(ESI) *m/z*, calcd for $C_{489}H_{718}N_{124}O_{139}S_3$: 10647.21; observed: $[M + 7H]^{7+}$ *m/z* = 1522.80, $[M + 8H]^{8+}$ *m/z* = 1332.80, $[M + 9H]^{9+}$ *m/z* = 1184.70, $[M + 10H]^{10+}$ *m/z* = 1066.30.

L-mGS2$^{38–90}$ (L-7). Peptides L-**2b** (5.4 mg) and L-**3** (5.2 mg) were dissolved in 1,2,4-triazole ligation buffer (1.1 mL), and the reaction mixture was incubated for 1 h at 37 °C. Then, 2 M $MeONH_2$ (0.12 mL) was added, and the reaction mixture was incubated for 2 h at 37 °C. After H_2O/CH_3CN (1:1) containing 0.1% (v/v) TFA was added, the mixture was purified by preparative HPLC (Cosmosil 5C18-AR300 column, a linear gradient of 28–58% CH_3CN in H_2O containing 0.1% (v/v) TFA over 90 min) to provide L-**7** (4.9 mg, 56% yield). MS(ESI) *m/z*, calcd for $C_{305}H_{437}N_{83}O_{84}S_3$: 6702.16; observed: $[M + 4H]^{4+}$ *m/z* = 1677.20, $[M + 5H]^{5+}$ *m/z* = 1342.38.

L-mGS2 (L-6, *C-to-N synthesis*). Peptides L-**1** (10.8 mg) and L-**7** (10.6 mg) were dissolved in 1,2,4-triazole ligation buffer (2.5 mL), and the reaction mixture was incubated for 1 h at 37 °C. After H_2O/CH_3CN (1:1) containing 0.1% (v/v) TFA was added, the mixture was purified by preparative HPLC (Cosmosil 5C18-AR300 column, a linear gradient of 30–60% CH_3CN in H_2O containing 0.1% (v/v) TFA over 90 min) to provide L-**6** (10.0 mg, 62% yield). MS(ESI) *m/z*, calcd for $C_{489}H_{718}N_{124}O_{139}S_3$: 10647.21; observed: $[M + 7H]^{7+}$ *m/z* = 1522.50, $[M + 8H]^{8+}$ *m/z* = 1332.29, $[M + 9H]^{9+}$ *m/z* = 1184.26, $[M + 10H]^{10+}$ *m/z* = 1066.22.

L-mGS2^{SS} (L-8). 1 mM 2-PDS in MeOH (0.93 mL) was added dropwise to a solution of L-mGS2 (L-**6**, 0.93 mg) in denaturation buffer [6 M guanidine, 20 mM Tris buffer (pH 8.0); 9.3 mL]. The reaction mixture was incubated for 1 h on ice. The mixture was purified by preparative HPLC (Cosmosil 5C4-AR300 column, a linear gradient of 30–60% CH_3CN in H_2O containing 0.1% (v/v) TFA over 90 min) to provide L-**8** (0.41 mg, 44% yield). MS(ESI) *m/z*, calcd for $C_{489}H_{716}N_{124}O_{139}S_3$: 10645.19; observed: $[M + 6]^{6+}$ *m/z* = 1775.55, $[M + 7H]^{7+}$ *m/z* = 1522.31, $[M + 8H]^{8+}$ *m/z* = 1332.11, $[M + 9H]^{9+}$ *m/z* = 1184.57.

D-mGS2$^{1-37}$ (D-1). By the identical procedure for the synthesis of peptide L-**1**, peptide D-**1** was synthesized (16.2 mg, 7% yield) from Fmoc-NH-SAL resin (70 mg, 0.040 mmol). MS(ESI) *m/z*, calcd for $C_{217}H_{338}N_{60}O_{61}$: 4760.52; observed: $[M + 3H]^{3+}$ *m/z* = 1589.09, $[M + 4H]^{4+}$ *m/z* = 1192.07, $[M + 5H]^{5+}$ *m/z* = 953.92.

[Thz^{38}]-D-mGS2$^{38-65}$ (D-2b). By the identical procedure for the synthesis of peptide L-**2b**, peptide D-**2b** was synthesized (30.3 mg, 24% yield) from Fmoc-NH-SAL resin (70 mg, 0.040 mmol). MS(ESI) *m/z*, calcd for $C_{145}H_{200}N_{36}O_{41}S$: 3133.44; observed: $[M + 2H]^{2+}$ *m/z* = 1568.76, $[M + 3H]^{3+}$ *m/z* = 1046.39.

D-mGS2$^{66-90}$ (D-3). By the identical procedure for the synthesis of peptide L-**3**, peptide D-**3** was synthesized (10.7 mg, 9% yield) from TGR resin (0.10 g, 0.025 mmol). MS(ESI) *m/z*, calcd for $C_{170}H_{246}N_{50}O_{45}S_2$: 3771.79; observed: $[M + 2H]^{2+}$ *m/z* = 1887.97, $[M + 3H]^{3+}$ *m/z* = 1258.58.

D-mGS2$^{38-90}$ (D-7). By the identical procedure for the synthesis of peptide L-**7**, peptide D-**7** was synthesized (5.6 mg, 59% yield) from D-**2b** (5.6 mg) and D-**3** (5.7 mg). MS(ESI) *m/z*, calcd for $C_{305}H_{437}N_{83}O_{84}S_3$: 6702.16; observed: $[M + 4H]^{4+}$ *m/z* = 1677.05, $[M + 5H]^{5+}$ *m/z* = 1341.67.

D-mGS2$^{1-90}$ (D-6). By the identical procedure for the synthesis of peptide L-**6**, peptide D-**6** was synthesized (4.3 mg, 63% yield) from D-**1** (5.0 mg) and D-**7** (5.3 mg). MS(ESI) *m/z*, calcd for $C_{489}H_{718}N_{124}O_{139}S_3$: 10647.21; observed: $[M + 8H]^{8+}$ *m/z* = 1332.68, $[M + 9H]^{9+}$ *m/z* = 1184.72, $[M + 10H]^{10+}$ *m/z* = 1065.99, $[M + 11H]^{11+}$ *m/z* = 968.93.

L-GS2$^{1-12}$ (L-9). By the same protocol for the synthesis of peptide L-**1**, peptide L-**9** was synthesized (50.1 mg, 23% yield) from Fmoc-NH SAL resin (0.14 g, 0.080 mmol). MS(ESI) *m/z*, calcd for $C_{88}H_{152}N_{32}O_{23}$: 2025.17; observed: $[M + 2H]^{2+}$ *m/z* = 1013.60, $[M + 3H]^{3+}$ *m/z* = 675.93.

[Cys^{13}]-L-GS2$^{13-48}$ (L-10). By the standard protocol for peptide synthesis, peptide L-**10** was synthesized (12.8 mg, 4% yield) from Fmoc-NH SAL resin (0.14 mg, 0.080 mmol). MS(ESI) *m/z*, calcd for $C_{232}H_{326}N_{62}O_{61}S$: 4988.40; observed: $[M + 3H]^{3+}$ *m/z* = 1663.85, $[M + 4H]^{4+}$ *m/z* = 1248.02, $[M + 5H]^{5+}$ *m/z* = 998.90.

[Thz^{49}]-L-GS2$^{49-73}$ (L-11). By the same protocol for the synthesis of peptide L-**1**, peptide L-**11** was synthesized (47 mg, 15% yield) from Fmoc-NH SAL resin (0.14 mg, 0.080 mmol). MS(ESI) *m/z*, calcd for $C_{145}H_{233}N_{45}O_{43}S$: 3324.71; observed: $[M + 3H]^{3+}$ *m/z* = 1109.68, $[M + 4H]^{4+}$ *m/z* = 832.36.

[Cys^{74}]-L-GS2$^{74–90/\text{biotin}}$ (L-12). By the standard protocol for peptide synthesis, peptide L-**12** was synthesized (30.0 mg, 9% yield) from TGR resin (0.32 g, 0.080 mmol). MS(ESI) *m/z*, calcd for $C_{146}H_{216}N_{48}O_{36}S_3$: 3313.57; observed: $[M + 2H]^{2+}$ *m/z* = 1657.55, $[M + 3H]^{3+}$ *m/z* = 1105.72.

[Cys(Trt-K$_{10}$)^{13}]-L-GS2$^{1–48}$ (L-13). Peptides L-**9** (8.5 mg) and L-**10** (9.6 mg) was dissolved in 1,2,4-triazole ligation buffer (3.0 mL), and the reaction mixture was incubated for 1 h at 37 °C. Then, Trt(OH)-K_{10} (25 mg) and TFA (3.0 mL) were added, and the reaction mixture was incubated for 1 h at room temperature. After H_2O/CH_3CN (1:1) containing 0.1% (v/v) TFA was added, the mixture was purified by preparative HPLC (Cosmosil 5C18-AR300 column, a linear gradient of 30–60% CH_3CN in H_2O containing 0.1% (v/v) TFA over 90 min) to provide L-**13** (6.6 mg, 40% yield). MS(ESI) *m/z*, calcd for $C_{371}H_{561}N_{97}O_{92}S$: 7879.19; observed: $[M + 5H]^{5+}$ *m/z* = 1576.86, $[M + 6H]^{6+}$ *m/z* = 1314.67, $[M + 7H]^{7+}$ *m/z* = 1126.94, $[M + 8H]^{8+}$ *m/z* = 986.16.

[Cys(Trt-K$_{10}$)^{13}]-L-GS2$^{1–48}$ MES ester (L-14). A $NaNO_2$ solution (60 μL) was added to a solution of L-**13** (6.1 mg) in activation buffer (0.60 mL) at − 20 °C, and the reaction was continued for 30 min. An MESNa solution [200 mM MESNa, 50 mM TCEP, 6 M guanidine, 200 mM phosphate buffer (pH 6.5); 0.20 mL] was then added to the reaction, and the reaction mixture was stirred for 5 min at − 20 °C. After H_2O/CH_3CN (1:1) containing 0.1% (v/v) TFA was added, the mixture was purified by preparative HPLC (Cosmosil 5C18-AR300 column, a linear gradient of 31–61% CH_3CN in H_2O containing 0.1% (v/v) TFA over 90 min) to provide L-**14** (3.4 mg, 64% yield). MS(ESI) *m/z*, calcd for $C_{342}H_{510}N_{78}O_{90}S_3$: 7245.69; observed: $[M + 4H]^{4+}$ *m/z* = 1813.12, $[M + 5H]^{5+}$ *m/z* = 1450.53, $[M + 6H]^{6+}$ *m/z* = 1209.21.

[Cys^{49,74}]-L-GS2$^{49–90/\text{biotin}}$ (L-15). By the same procedure for the synthesis of peptide L-**7**, peptide L-**15** was synthesized (5.0 mg, 29% yield) from L-**11** (13.4 mg) and L-**12** (10.8 mg). MS(ESI) *m/z*, calcd for $C_{257}H_{392}N_{74}O_{73}S_4$: 5810.81; observed: $[M + 4H]^{4+}$ *m/z* = 1454.07, $[M + 5H]^{5+}$ *m/z* = 1163.24.

[Cys(Trt-K$_{10}$)^{13}, Cys^{49,74}]- L-GS2^{biotin} (L-16). Peptides L-**14** (3.4 mg) and L-**15** (2.5 mg) was dissolved in MPAA ligation buffer [100 mM MPAA, 50 mM TCEP, 6 M guanidine, 200 mM phosphate (pH 6.5); 0.72 mL], and the reaction mixture was incubated for 1 h at 37 °C. After H_2O/CH_3CN (1:1) containing 0.1% (v/v) TFA was added, the mixture was purified by preparative HPLC (Cosmosil 5C18-AR300 column, a linear gradient of 33–63% CH_3CN in H_2O containing 0.1% (v/v) TFA over 90 min) to provide L-**16** (2.4 mg, 43% yield). MS(ESI) *m/z*, calcd for $C_{597}H_{896}N_{152}O_{160}S_5$: 12914.53; observed: $[M + 7H]^{7+}$ *m/z* = 1846.18, $[M + 8H]^{8+}$ *m/z* = 1615.37, $[M + 9H]^{9+}$ *m/z* = 1436.01, $[M + 10H]^{10+}$ *m/z* = 1292.91, $[M + 11H]^{11+}$ *m/z* = 1175.18, $[M + 12H]^{12+}$ *m/z* = 1077.62.

[Cys^{13,49,74}]-L-GS2^{biotin} (L-17). Peptide L-**16** (2.1 mg) was dissolved in TFA/EDT (95/5) and the solution was incubated for 30 min at room temperature. The reaction mixture was poured into ice-cold dry Et_2O. The precipitate was dissolved in 6 M guanidine solution (0.20 mL), and purified by preparative HPLC (Cosmosil

5C4-AR300 column, a linear gradient of 31–61% CH_3CN in H_2O containing 0.1% (v/v) TFA over 90 min) to provide L-**17** (1.0 mg, 60%). MS(ESI) *m/z*, calcd for $C_{513}H_{756}N_{130}O_{146}S_5$: 11233.43; observed: $[M + 7H]^{7+}$ *m/z* = 1606.33, $[M + 8H]^{8+}$ *m/z* = 1406.13, $[M + 9H]^{9+}$ *m/z* = 1250.30, $[M + 10H]^{10+}$ *m/z* = 1124.78.

L-GS2^{biotin} (L-18). L-**17** (1.0 mg) was dissolved in desulfurization buffer [20 mM VA-044, 100 mM MESNa, 250 mM TCEP, 6 M guanidine, 100 mM phosphate (pH 6.5); 0.42 mL], and incubated for 2 h at 37 °C. After H_2O/CH_3CN (1:1) containing 0.1% (v/v) TFA was added, the mixture was purified by preparative HPLC (Cosmosil 5C4-AR300 column, a linear gradient of 31–61% CH_3CN in H_2O containing 0.1% (v/v) TFA over 90 min) to provide L-**18** (0.53 mg, 51%). MS(ESI) *m/z*, calcd for $C_{513}H_{756}N_{130}O_{146}S_2$: 11137.52; observed: $[M + 7H]^{7+}$ *m/z* = 1592.81, $[M + 8H]^{8+}$ *m/z* = 1393.76, $[M + 9H]^{9+}$ *m/z* = 1239.10, $[M + 10H]^{10+}$ *m/z* = 1115.06.

[Cys(Acm)^{66}]-L-mGS2$^{66–73}$ (L-19). By the same protocol for the synthesis of peptide L-**1**, peptide L-**19** was synthesized (92.0 mg, 48% yield) from Fmoc-NH SAL resin (0.14 g, 0.080 mmol). MS(ESI) *m/z*, calcd for $C_{80}H_{124}N_{28}O_{21}S$: 1844.92; observed: $[M + 2H]^{2+}$ *m/z* = 923.64, $[M + 3H]^{3+}$ *m/z* = 615.95.

[Cys(Acm)^{66}]-L-mGS2$^{66–90/biotin}$ (L-20). Peptides L-**19** (17.1 mg) and L-**12** (19.4 mg) were dissolved in 1,2,4-triazole ligation buffer (4.5 mL), and the reaction mixture was incubated for 1 h at 37 °C. Then, desulfurization buffer [450 mM TCEP, 80 mM MESNa, 6 M guanidine, 100 mM phosphate (pH 4.5); 4.5 mL] and 1 M VA-044 (0.45 mL) were added, and the pH was adjusted to 6.5. The reaction mixture was further incubated for 2 h at 37 °C. The product was purified by preparative HPLC (Cosmosil 5C18-AR300 column, a linear gradient of 27–57% CH_3CN in H_2O containing 0.1% (v/v) TFA over 90 min) to provide L-**20** (8.4 mg, 35% yield). MS(ESI) *m/z*, calcd for $C_{193}H_{283}N_{57}O_{51}S_3$: 4311.05; observed: $[M + 3H]^{3+}$ *m/z* = 1438.56, $[M + 4H]^{4+}$ *m/z* = 1079.05, $[M + 5H]^{5+}$ *m/z* = 863.53.

L-mGS2$^{66–90/biotin}$ (L-21). Peptide L-**20** (7.4 mg) was dissolved in deprotection cocktail [AgOTf (27 mg), anisole (10 μL), TFA (0.50 mL)], and the reaction mixture was incubated for 1.5 h on ice. The product was poured into ice-cold Et_2O, and the precipitate was collected by centrifugation. The precipitate was washed by reduction buffer [10 mM DTT, 6 M guanidine, 100 mM phosphate (pH 7.4)], and the supernatant was purified by preparative HPLC (Cosmosil 5C18-AR300 column, a linear gradient of 27–57% CH_3CN in H_2O containing 0.1% (v/v) TFA over 90 min) to provide L-**21** (5.0 mg, 69%). MS(ESI) *m/z*, calcd for $C_{190}H_{278}N_{56}O_{50}S_3$: 4240.01; observed: $[M + 3H]^{3+}$ *m/z* = 1414.80, $[M + 4H]^{4+}$ *m/z* = 1061.37.

L-mGS2$^{38–90/biotin}$ (L-22). By the same procedure for the synthesis of peptide L-**7**, peptide L-**22** was synthesized (3.2 mg, 44% yield) from L-**2b** (4.9 mg) and L-**21** (4.6 mg). MS(ESI) *m/z*, calcd for $C_{325}H_{469}N_{89}O_{89}S_4$: 7170.38; observed: $[M + 4H]^{4+}$ *m/z* = 1793.93, $[M + 5H]^{5+}$ *m/z* = 1435.37, $[M + 6H]^{6+}$ *m/z* = 1196.63.

L-mGS2^{biotin} (L-23). By the same procedure for the synthesis of peptide L-**6**, peptide L-**23** was synthesized (1.4 mg, 31% yield) from L-**1** (3.3 mg) and L-**22** (3.0 mg). MS(ESI) *m/z*, calcd for $C_{509}H_{750}N_{130}O_{144}S_4$: 11115.42; observed: $[M + 8H]^{8+}$ *m/z*

= 1390.97, $[M + 9H]^{9+}$ *m/z* = 1236.70, $[M + 10H]^{10+}$ *m/z* = 1113.26, $[M + 11H]^{11+}$ *m/z* = 1012.18.

L-mGS2$^{SS/biotin}$ (L-24). By the same procedure for the synthesis of peptide L-**8**, peptide L-**24** was synthesized (0.08 mg, 9% yield) from L-**23** (0.93 mg). MS(ESI) *m/z*, calcd for $C_{509}H_{748}N_{130}O_{144}S_4$: 11113.41; observed: $[M + 8H]^{8+}$ *m/z* = 1390.48, $[M + 9H]^{9+}$ *m/z* = 1237.23, $[M + 10H]^{10+}$ *m/z* = 1113.31, $[M + 11H]^{11+}$ *m/z* = 1012.01.

[Cys(Acm)^{66}]-D-mGS2$^{66–73}$ (D-19). By the identical procedure for the synthesis of peptide L-**19**, peptide D-**19** was synthesized (63.1 mg, 32% yield) from Fmoc-NH SAL resin (0.14 g, 0.080 mmol). MS(ESI) *m/z*, calcd for $C_{80}H_{124}N_{28}O_{21}S$: 1844.92; observed: $[M + 2H]^{2+}$ *m/z* = 923.66, $[M + 3H]^{3+}$ *m/z* = 615.96.

D-mGS2$^{74–90/biotin}$ (D-12). By the identical procedure for the synthesis of peptide L-**12**, peptide D-**12** was synthesized (24.4 mg, 7% yield) from TGR resin (0.32 g, 0.080 mmol). MS(ESI) *m/z*, calcd for $C_{146}H_{216}N_{48}O_{36}S_3$: 3313.57; observed: $[M + 3H]^{3+}$ *m/z* = 1106.81, $[M + 4H]^{4+}$ *m/z* = 830.17.

[Cys(Acm)^{66}]-D-mGS2$^{66–90/biotin}$ (D-20). By the identical procedure for the synthesis of peptide L-**20**, peptide D-**20** was synthesized (18.0 mg, 63% yield) from D-**19** (19.5 mg) and D-**12** (23.2 mg). MS(ESI) *m/z*, calcd for $C_{193}H_{283}N_{57}O_{51}S_3$: 4311.05; observed: $[M + 3H]^{3+}$ *m/z* = 1439.09, $[M + 4H]^{4+}$ *m/z* = 1079.72, $[M + 5H]^{5+}$ *m/z* = 863.88.

D-mGS2$^{66–90/biotin}$ (D-21). By the identical procedure for the synthesis of peptide L-**21**, peptide D-**21** was synthesized (2.2 mg, 22% yield) from D-**20** (10.4 mg). MS(ESI) *m/z*, calcd for $C_{190}H_{278}N_{56}O_{50}S_3$: 4240.01; observed: $[M + 3H]^{3+}$ *m/z* = 1415.72, $[M + 4H]^{4+}$ *m/z* = 1061.76.

D-mGS2$^{38–90/biotin}$ (D-22). By the identical procedure for the synthesis of peptide L-**22**, peptide D-**22** was synthesized (2.1 mg, 61% yield) from D-**2b** (2.2 mg) and D-**21** (2.2 mg). MS(ESI) *m/z*, calcd for $C_{325}H_{469}N_{89}O_{89}S_4$: 7170.38; observed: $[M + 5H]^{5+}$ *m/z* = 1436.29, $[M + 6H]^{6+}$ *m/z* = 1196.82, $[M + 7H]^{7+}$ *m/z* = 1026.18.

D-mGS2^{biotin} (D-23). By the identical procedure for the synthesis of peptide L-**23**, peptide D-**23** was synthesized (1.3 mg, 45% yield) from D-**1** (2.0 mg) and D-**22** (2.0 mg). MS(ESI) *m/z*, calcd for $C_{509}H_{750}N_{130}O_{144}S_4$: 11115.42; observed: $[M + 8H]^{8+}$ *m/z* = 1391.50, $[M + 10H]^{10+}$ *m/z* = 1113.37, $[M + 11H]^{11+}$ *m/z* = 1012.46.

Folding of L/D-mGS2 and L/D-mGS2^{biotin}. The lyophilized peptide was dissolved in guanidine solution [5 mM TCEP, 6 M guanidine, 20 mM Tris (pH 8); 0.3 mg/mL], and incubated for 1 h at room temperature. The denatured protein was subjected to dialysis using Slide-A-Lyzer G2 dialysis cassette (cutoff 3.5 kDa, Thermo) against a 200-fold volume of dialysis buffer [10 mM TCEP, 50 mM acetate buffer (pH 4.5)] for 1.5 h and 12 h at 4 °C. The folded proteins were used for to the following experiments without purification.

Folding of L-mGS2^{SS} and L-mGS2$^{SS/biotin}$. The lyophilized peptide was dissolved in guanidine solution [6 M guanidine, 20 mM Tris (pH 8); 0.3 mg/mL], and incubated for 1 h at room temperature. The denatured protein was subjected to dialysis using Slide-A-Lyzer G2 dialysis cassette (cutoff 3.5 kDa, Thermo) against a 200-fold volume of dialysis buffer [50 mM acetate buffer (pH 4.5)] for 1.5 h and 12 h at 4 °C. The folded protein was used for the following experiments without purification.

Folding of L-GS2^{biotin}. The lyophilized peptide was dissolved in guanidine solution [6 M guanidine, 20 mM Tris (pH 8); 0.3 mg/mL], and incubated for 1 h at room temperature. The denatured protein was subjected to dialysis using Slide-A-Lyzer G2 dialysis cassette (cutoff 3.5 kDa, Thermo) against a 200-fold volume of dialysis buffer [50 mM acetate buffer (pH 4.5)] for 1.5 h and 12 h at 4 °C. The folded protein was used for the following experiments without purification.

CD Spectra Measurement. Refolded protein in acetate buffer [50 mM acetate buffer (pH 4.5)] was used for measurement. CD spectrum was recorded on a J-720 circular dichroism spectrometer (JASCO) at 20 °C. The thermal denaturation was measured by observing the transition of CD signal at 203 nm during heating at a rate of 1 °C/min (protein concentration: 20 μM).

SPR Analysis. SPR analysis of monobody proteins binding to GST-EGFP was carried out using a Biacore X-100 (Cytiva). After the immobilization of biotinylated monobody proteins on a sensor chip SA, the analyte protein solution of GST-EGFP were flashed into the flow system in the running buffer [10 mM HEPES (pH 7.4), 150 mM NaCl, 3 mM EDTA, 0.005% Surfactant P20] at 25 °C. All analytes were evaluated for 90 s as contact time, followed by 600 s dissociation at a flow rate of 50 μL/min. The triplicate data were analyzed.

Immunogenicity Assays. All animal experiments were carried out in accordance with the author's institutional guidelines and approved by the Kyoto University Animal Care Committee. BALB/c mice (female, 5 weeks old) were given intraperitoneal injections on day 0 with 50 μg/injection of synthetic L-mGS2 or D-mGS2 antigen emulsified in Freund's Complete Adjuvant (Wako) and on days 14 and 28 with 50 μg/injection of synthetic L-mGS2 or D-mGS2 antigen emulsified in Freund's Incomplete Adjuvant (Wako). Immune plasma samples were collected on day 0, 14, 28, 35 and 44.

ELISAs were performed in PBS (pH 7.4) containing 0.05% Tween 20 for all wash and dilution processes. 96-well Maxisorp microtiter plates (Thermo) were coated overnight at 4 °C with either L-mGS2 or D-mGS2 in 50 mM acetate buffer (pH 4.4) (50 μL/well; 1000 ng/mL). After coating, wells were washed three times and blocked with PBS containing 3% BSA (150 μL/well) for 2 h. After three times washes, 1:1,000 dilution of immunized plasma from each mouse (50 μL/well; day 0, 14, 28, 35 and 44) or serial dilution of immunized plasma from each mouse (50 μL/well; day 44) was added and incubated for 1 h. After three times washes, 1:10,000 dilution of anti-mouse IgG (H + L), HRP conjugate (Jackson ImmunoResearch Laboratories) (50 μL/well) was added and incubated for 30 min. After five times washes, TMB solution was added and incubated for 2 min, then the reaction was stopped using 1 M H_2SO_4 (50 μL/well). Absorbance was measured at 450 nm using an infinite M200 Plate Reader (TECAN).

Anti-D-MCP-1 Monobody

Fmoc-NHNH-Trt(2-Cl) resin. Fmoc-NHNH-Trt(2-Cl) resin was prepared according to the previously reported procedure [34]. A solution of Fmoc-$NHNH_2$ (321 mg, 1.24 mmol) and DIEA (1.06 mL, 6.06 mmol) in dry DMF/CH_2Cl_2 (4:1) was reacted with Trt(2-Cl) chloride resin (2.80 g, 3.03 mmol) for 3 h at room temperature.

L-mMb5-11A$^{1–39}$ MPAA thioester (L-25a). Tyr39 was manually loaded on Fmoc-NHNH-Trt(2-Cl) resin (0.25 g, 0.10 mmol) by Fmoc-Tyr(*t*-Bu)-OH (0.23 g, 0.50 mmol), OxymaPure (71 mg, 0.50 mmol), and DIC (78 μL, 0.50 mmol) in DMF. The remaining peptide sequence was constructed using a microwave-assisted peptide synthesizer. The resin was then treated with TFA/H_2O/*m*-cresol/thioanisole/EDT (80:5:5:5:5) for 2 h at room temperature. After removal of the resin by filtration, the filtrate was poured into ice-cold dry Et_2O. The precipitate was dissolved in MPAA suspension (200 mM MPAA, 6 M guanidine; 5.0 mL). Then, acac (51 μL, 0.50 mmol) was added and the reaction was continued for 2 h at room temperature. After H_2O/CH_3CN (1:1) containing 0.05% (v/v) TFA (3.3 mL) was added, the mixture was purified by preparative HPLC (Cosmosil Protein-R column, a linear gradient of 23–53% CH_3CN in H_2O containing 0.05% (v/v) TFA over 90 min) to provide L-**25a** (23.9 mg, 4% yield from resin). MS(ESI) *m/z*, calcd for $C_{210}H_{298}N_{56}O_{61}S_2$: 4644.14; observed: $[M + 3H]^{3+}$ *m/z* = 1549.85, $[M + 4H]^{4+}$ *m/z* = 1162.60, $[M + 5H]^{5+}$ *m/z* = 930.25.

[Thz^{40}]-L-mMb5-11A$^{40–79}$ MPAA thioester (L-26a). By the same protocol for the synthesis of peptide L-**25a**, peptide L-**26a** was synthesized (56.6 mg, 13% yield) from Fmoc-NHNH-Trt(2-Cl) resin (0.24 g, 0.10 mmol). MS(ESI) *m/z*, calcd for $C_{188}H_{290}N_{44}O_{63}S_2$: 4239.04; observed: $[M + 3H]^{3+}$ *m/z* = 1413.70.

L-mMb5-11A$^{80–109}$ (L-27a). Peptide L-**27a** was synthesized (29.0 mg, 5% yield) from TGR resin (0.40 g, 0.10 mmol) using a microwave-assisted peptide synthesizer. MS(ESI) *m/z*, calcd for $C_{195}H_{277}N_{61}O_{55}S$: 4385.05; observed: $[M + 3H]^{3+}$ *m/z* = 1463.40, $[M + 4H]^{4+}$ *m/z* = 1097.85.

L-mMb5-11A$^{40–109}$ (L-28a). Peptides L-**26a** (33.5 mg) and L-**27a** (33.4 mg) were dissolved in MPAA ligation buffer with increased TCEP concentration [100 mM MPAA, 100 mM TCEP, 6 M guanidine, 100 mM phosphate (pH 7.0); 6.0 mL], and the reaction mixture was incubated for 15 h at 37 °C. Then, TCEP (86 mg) and 2 M $MeONH_2$ (0.67 mL) were added, and the reaction mixture was further incubated for 2 h at 37 °C. The mixture was purified by preparative HPLC (Cosmosil Protein-R column, a linear gradient of 23–53% CH_3CN in H_2O containing 0.05% (v/v) TFA over 90 min) to provide L-**28a** (42.6 mg, 72% yield). MS(ESI) *m/z*, calcd for $C_{374}H_{559}N_{105}O_{116}S_2$: 8441.05; observed: $[M + 5H]^{5+}$ *m/z* = 1690.10, $[M + 6H]^{6+}$ *m/z* = 1408.75, $[M + 7H]^{7+}$ *m/z* = 1207.45, $[M + 8H]^{8+}$ *m/z* = 1056.75.

L-mMb5-11A (L-29a). Peptides L-**25a** (23.9 mg) and L-**28a** (42.6 mg) were dissolved in MPAA ligation buffer [100 mM MPAA, 50 mM TCEP, 6 M guanidine, 100 mM phosphate (pH 7.0); 4.5 mL], and the reaction mixture was incubated for 1 h at 37 °C. The mixture was purified by preparative HPLC (Cosmosil Protein-R column, a linear gradient of 27–57% CH_3CN in H_2O containing 0.05% (v/v) TFA over 90 min) to provide L-**29a** (21.0 mg, 32%). MS(ESI) *m/z*, calcd for $C_{576}H_{849}N_{161}O_{175}S_3$: 12917.16; observed: $[M + 8H]^{8+}$ *m/z* = 1616.25, $[M + 9H]^{9+}$ *m/z* = 1436.70, $[M + 10H]^{10+}$ *m/z* = 1293.35, $[M + 11H]^{11+}$ *m/z* = 1175.80, $[M + 12H]^{12+}$ *m/z* = 1077.85, $[M + 13H]^{13+}$ *m/z* = 995.00.

L-mMb5-11B$^{1–41}$ MPAA thioester (L-25b). By the same protocol for the synthesis of peptide L-**25a**, peptide L-**25b** was synthesized (8.8 mg, 3% yield) from Fmoc-NHNH-Trt(2-Cl) resin (0.25 g, 0.10 mmol). MS(ESI) *m/z*, calcd for $C_{222}H_{321}N_{61}O_{63}S_2$: 4913.32; observed: $[M + 3H]^{3+}$ *m/z* = 1639.80, $[M + 4H]^{4+}$ *m/z* = 1230.10.

[Thz^{42}]-L-mMb5-11B$^{42–77}$ MPAA thioester (L-26b). By the same protocol for the synthesis of peptide L-**25a**, peptide L-**26b** was synthesized (51.3 mg, 12% yield) from Fmoc-NHNH-Trt(2-Cl) resin (0.25 g, 0.10 mmol). MS(ESI) *m/z*, calcd for $C_{169}H_{256}N_{40}O_{57}S_2$: 3821.78; observed: $[M + 3H]^{3+}$ *m/z* = 1275.55.

L-mMb5-11B$^{78–109}$ (L-27b). Peptide L-**27b** was synthesized (20.4 mg, 4% yield) from TGR resin (0.40 g, 0.10 mmol) using a microwave-assisted peptide synthesizer. MS(ESI) *m/z*, calcd for $C_{209}H_{295}N_{63}O_{58}S$: 4647.18; observed: $[M + 3H]^{3+}$ *m/z* = 1550.90, $[M + 4H]^{4+}$ *m/z* = 1163.40.

L-mMb5-11B$^{42–109}$ (L-28b). By the same protocol for the synthesis of peptide L-**28a**, peptide L-**28b** was synthesized (12.4 mg, 39% yield) from peptides L-**26b** (20.4 mg) and L-**27b** (18.9 mg). MS(ESI) *m/z*, calcd for $C_{369}H_{543}N_{103}O_{113}S_2$: 8288.94; observed: $[M + 5H]^{5+}$ *m/z* = 1659.70, $[M + 6H]^{6+}$ *m/z* = 1383.50, $[M + 7H]^{7+}$ *m/z* = 1185.80, $[M + 8H]^{8+}$ *m/z* = 1037.70.

L-mMb5-11B (L-29b). Peptides L-**25b** (4.4 mg) and L-**28b** (4.9 mg) were dissolved in MPAA ligation buffer with increased TCEP concentration (0.50 mL), and the reaction mixture was incubated for 22 h at 37 °C. The mixture was purified by preparative HPLC (Cosmosil Protein-R column, a linear gradient of 25–55% CH_3CN in H_2O containing 0.05% (v/v) TFA over 90 min) to provide L-**29b** (3.6 mg, 48%). MS(ESI) *m/z*, calcd for $C_{583}H_{856}N_{164}O_{174}S_3$: 13034.23; observed: $[M + 8H]^{8+}$ *m/z* = 1631.35, $[M + 9H]^{9+}$ *m/z* = 1450.20, $[M + 10H]^{10+}$ *m/z* = 1305.25, $[M + 11H]^{11+}$ *m/z* = 1186.65, $[M + 13H]^{13+}$ *m/z* = 1004.25.

L-mMb5-11C$^{1–44}$ MPAA thioester (L-25c). By the same protocol for the synthesis of peptide L-**25a**, peptide L-**25c** was synthesized (12.6 mg, 4% yield) from Fmoc-NHNH-Trt(2-Cl) resin (0.13 g, 0.050 mmol). MS(ESI) *m/z*, calcd for $C_{237}H_{340}N_{64}O_{68}S_2$: 5234.46; observed: $[M + 3H]^{3+}$ *m/z* = 1746.75, $[M + 4H]^{4+}$ *m/z* = 1310.20, $[M + 5H]^{5+}$ *m/z* = 1048.40.

[Thz^{45}]-L-mMb5-11C$^{45–72}$ MPAA thioester (L-26c). By the same protocol for the synthesis of peptide L-**25a**, peptide L-**26c** was synthesized (81.5 mg, 25% yield) from Fmoc-NHNH-Trt(2-Cl) resin (0.25 g, 0.10 mmol). MS(ESI) *m/z*, calcd for

$C_{125}H_{196}N_{32}O_{42}S_2$: 2881.36; observed: $[M + 2H]^{2+}$ *m/z* = 1442.30, $[M + 3H]^{3+}$ *m/z* = 961.85.

L-mMb5-11C$^{73-109}$ (L-27c). Peptide L-**27c** was synthesized (54.3 mg, 4% yield) from TGR resin (0.80 g, 0.20 mmol) using a microwave-assisted peptide synthesizer. MS(ESI) *m/z*, calcd for $C_{236}H_{334}N_{68}O_{68}S$: 5240.45; observed: $[M + 3H]^{3+}$ *m/z* = 1748.70, $[M + 4H]^{4+}$ *m/z* = 1311.80, $[M + 5H]^{5+}$ *m/z* = 1049.65.

L-mMb5-11C$^{45-109}$ (L-28c). Peptides L-**26c** (838 mg) and L-**27c** (14.1 mg) were dissolved in MPAA ligation buffer (3.2 mL), and the reaction mixture was incubated for 1 h at 37 °C. Then, 2 M $MeONH_2$ (0.24 mL) were added, and the reaction mixture was further incubated for 2 h at 37 °C. The mixture was purified by preparative HPLC (Cosmosil Protein-R column, a linear gradient of 23–53% CH_3CN in H_2O containing 0.05% (v/v) TFA over 90 min) to provide L-**28c** (13.3 mg, 64% yield). MS(ESI) *m/z*, calcd for $C_{352}H_{522}N_{100}O_{108}S_2$: 7941.79; observed: $[M + 5H]^{5+}$ *m/z* = 1590.15, $[M + 6H]^{6+}$ *m/z* = 1325.25, $[M + 7H]^{7+}$ *m/z* = 1136.00, $[M + 8H]^{8+}$ *m/z* = 994.25.

L-mMb5-11C (L-29c). By the same protocol for the synthesis of peptide L-**29a**, peptide L-**29c** was synthesized (25.2 mg, 68% yield) from peptides L-**25c** (15.3 mg) and L-**28c** (23.1 mg). MS(ESI) *m/z*, calcd for $C_{581}H_{854}N_{164}O_{174}S_3$: 13008.22; observed: $[M + 8H]^{8+}$ *m/z* = 1627.85, $[M + 9H]^{9+}$ *m/z* = 1447.15, $[M + 10H]^{10+}$ *m/z* = 1302.45, $[M + 11H]^{11+}$ *m/z* = 1184.15, $[M + 12H]^{12+}$ *m/z* = 1085.55, $[M + 13H]^{13+}$ *m/z* = 1002.10.

D-mMb5-11C$^{1-44}$ MPAA thioester (D-25c). By the identical procedure for the synthesis of peptide L-**25c**, peptide D-**25c** was synthesized (15.0 mg, 3% yield) from Fmoc-NHNH-Trt(2-Cl) resin (0.25 g, 0.10 mmol). MS(ESI) *m/z*, calcd for $C_{237}H_{340}N_{64}O_{68}S_2$: 5234.46; observed: $[M + 3H]^{3+}$ *m/z* = 1746.20, $[M + 4H]^{4+}$ *m/z* = 1309.85, $[M + 5H]^{5+}$ *m/z* = 1048.05.

[Thz^{45}]-D-mMb5-11C$^{45-72}$ MPAA thioester (D-26c). By the identical procedure for the synthesis of peptide L-**26c**, peptide D-**26c** was synthesized (57.9 mg, 18% yield) from Fmoc-NHNH-Trt(2-Cl) resin (0.25 g, 0.10 mmol). MS(ESI) *m/z*, calcd for $C_{125}H_{196}N_{32}O_{42}S_2$: 2881.36; observed: $[M + 2H]^{2+}$ *m/z* = 1442.05, $[M + 3H]^{3+}$ *m/z* = 961.70.

D-mMb5-11C$^{73-109}$ (D-27c). By the identical procedure for the synthesis of peptide L-**27c**, peptide D-**27c** was synthesized (30.8 mg, 5% yield) from TGR resin (0.40 g, 0.10 mmol). MS(ESI) *m/z*, calcd for $C_{236}H_{334}N_{68}O_{68}S$: 5240.45; observed: $[M + 3H]^{3+}$ *m/z* = 1748.55, $[M + 4H]^{4+}$ *m/z* = 1311.55, $[M + 5H]^{5+}$ *m/z* = 1049.40.

D-mMb5-11C$^{45-109}$ (D-28c). By the identical procedure for the synthesis of peptide L-**28c**, peptide D-**28c** was synthesized (14.3 mg, 45% yield) from peptides D-**26c** (13.3 mg) and D-**27c** (21.2 mg). MS(ESI) *m/z*, calcd for $C_{352}H_{522}N_{100}O_{108}S_2$: 7941.79; observed: $[M + 5H]^{5+}$ *m/z* = 1589.60, $[M + 6H]^{6+}$ *m/z* = 1324.95, $[M + 7H]^{7+}$ *m/z* = 1135.75, $[M + 8H]^{8+}$ *m/z* = 993.80.

D-mMb5-11C (D-29c). By the identical procedure for the synthesis of peptide L-**29c**, peptide D-**29c** was synthesized (15.6 mg, 70% yield) from peptides D-**25c** (9.1 mg) and D-**26c** (13.7 mg). MS(ESI) *m/z*, calcd for $C_{581}H_{854}N_{164}O_{185}S_3$: 13008.22; observed: $[M + 8H]^{2+}$ *m/z* = 1627.85, $[M + 9H]^{9+}$ *m/z* = 1447.15, $[M + 10H]^{10+}$ *m/z* = 1302.60, $[M + 11H]^{11+}$ *m/z* = 1184.30, $[M + 12H]^{12+}$ *m/z* = 1085.70, $[M + 13H]^{13+}$ *m/z* = 1002.15.

iNoc-L-Mb5-11$^{1-44}$ MPAA thioester (L-30). Gly44 was manually loaded on Fmoc-NHNH-Trt(2-Cl) resin (0.25 g, 0.10 mmol) by Fmoc-Gly-OH (0.15 g, 0.50 mmol), OxymaPure (71 mg, 0.50 mmol), and DIC (78 μL, 0.50 mmol) in DMF. The remaining peptide sequence was constructed using a microwave-assisted peptide synthesizer. The N-terminal amine was protected with iNoc group by 4-nitrophenyl-4-pyridylmethyl carbonate [37] (82 mg, 0.30 mmol), DIEA (0.10 mL, 0.60 mmol) in DMF for 2 h at room temperature. Half of the resin (0.050 mmol) was then treated with TFA/H_2O/*m*-cresol/thioanisole/EDT (80:5:5:5:5) for 2 h at room temperature. After removal of the resin by filtration, the filtrate was poured into ice-cold dry Et_2O. The precipitate was dissolved in MPAA suspension (200 mM MPAA, 6 M guanidine; 2.5 mL). Then, acac (26 μL, 0.25 mmol) was added and the reaction was continued for 1 h at room temperature. After H_2O/CH_3CN (1:1) containing 0.05% (v/v) TFA (1.7 mL) was added, the mixture was purified by preparative HPLC (Cosmosil Protein-R column, a linear gradient of 23–53% CH_3CN in H_2O containing 0.05% (v/v) TFA over 90 min) to provide L-**30** (9.6 mg, 3% yield from resin). MS(ESI) *m/z*, calcd for $C_{244}H_{345}N_{65}O_{70}S_2$: 5369.49; observed: $[M + 3H]^{3+}$ *m/z* = 1791.90, $[M + 4H]^{4+}$ *m/z* = 1344.10, $[M + 5H]^{5+}$ *m/z* = 1075.40.

Fmoc-[Lys(iNoc)^{61,70}]-L-Mb5-11$^{45-72}$ MPAA thioester (L-31). Gly72 and Pro71 were manually loaded on Fmoc-NHNH-Trt(2-Cl) resin (0.25 g, 0.10 mmol) by Fmoc-Gly-OH (0.15 g, 0.50 mmol) or Fmoc-Pro-OH (0.17 g, 0.50 mmol), OxymaPure (71 mg, 0.50 mmol), and DIC (78 μL, 0.50 mmol) in DMF. Lys70 and Lys61 were manually coupled by Fmoc-Lys(iNoc)-OH [51] (0.15 g, 0.30 mmol), OxymaPure (42 mg, 0.30 mmol), and DIC (46 μL, 0.30 mmol) in DMF for 2 h at room temperature. The remaining peptide sequence was constructed using a microwave-assisted peptide synthesizer. Fmoc group at the N-terminal amine was not deprotected. The resin was then treated with TFA/H_2O/*m*-cresol/thioanisole/EDT (80:5:5:5:5) for 2 h at room temperature. After removal of the resin by filtration, the filtrate was poured into ice-cold dry Et_2O. The precipitate was dissolved in MPAA suspension (200 mM MPAA, 6 M guanidine; 5.0 mL). Then, acac (51 μL, 0.50 mmol) was added and the reaction was continued for 1 h at room temperature. After H_2O/CH_3CN (1:1) containing 0.05% (v/v) TFA (3.3 mL) was added, the mixture was purified by preparative HPLC (Cosmosil Protein-R column, a linear gradient of 29–59% CH_3CN in H_2O containing 0.05% (v/v) TFA over 90 min) to provide L-**31** (9.3 mg, 3% yield from resin). MS(ESI) *m/z*, calcd for $C_{155}H_{218}N_{34}O_{50}S$: 3387.53; observed: $[M + 2H]^{2+}$ *m/z* = 1695.45, $[M + 3H]^{3+}$ *m/z* = 1130.60.

[Lys(iNoc)^{106}]-L-Mb5-11$^{73-109}$ (L-32). Peptide L-**32** was synthesized (21.3 mg, 3% yield) from TGR resin (0.40 g, 0.10 mmol) using a microwave-assisted peptide synthesizer. Lys106 was manually coupled by Fmoc-Lys(iNoc)-OH (0.15 g, 0.30 mmol), OxymaPure (42 mg, 0.30 mmol), and DIC (46 μL, 0.30 mmol) in DMF for 2 h at room temperature. MS(ESI) *m/z*, calcd for $C_{245}H_{343}N_{69}O_{70}$: 5371.54; observed: $[M + 3H]^{3+}$ *m/z* = 1792.70, $[M + 4H]^{4+}$ *m/z* = 1344.75, $[M + 5H]^{5+}$ *m/z* = 1076.00.

[Lys(iNoc)^{61,70,106}]-L-Mb5-11$^{45–109}$ (L-33). Peptides L-**31** (8.6 mg) and L-**32** (10.8 mg) were dissolved in coupling solution (150 mM HOOBt, 100 mM DIEA in DMSO; 0.33 mL), and the reaction mixture was incubated for 6 h at room temperature. Then, 20% piperidine in DMSO (0.33 mL) was added, and the reaction mixture was further incubated for 5 min at room temperature. The mixture was poured into ice-cold dry Et_2O, and the precipitate was purified by preparative HPLC (Cosmosil 5C18-AR300 column, a linear gradient of 25–55% CH_3CN in H_2O containing 0.05% (v/v) TFA over 90 min) to provide L-**33** (9.1 mg, 56% yield). MS(ESI) *m/z*, calcd for $C_{377}H_{543}N_{103}O_{116}$: 8368.98; observed: $[M + 5H]^{5+}$ *m/z* = 1675.65, $[M + 6H]^{6+}$ *m/z* = 1396.40, $[M + 7H]^{7+}$ *m/z* = 1197.10, $[M + 8H]^{8+}$ *m/z* = 1047.60.

iNoc-[Lys(iNoc)^{61,70,106}]-L-Mb5-11 (L-34). Peptides L-**30** (7.2 mg) and L-**33** (4.4 mg) were dissolved in coupling solution (90 μL), and the reaction mixture was incubated for 10 h at room temperature. The mixture was poured into EtOAc/Et_2O (1:1), and the precipitate was purified by preparative HPLC (Cosmosil 5C18-AR300 column, a linear gradient of 25–55% CH_3CN in H_2O containing 0.05% (v/v) TFA over 90 min) to provide L-**34** (4.1 mg, 58% yield). MS(ESI) *m/z*, calcd for $C_{613}H_{880}N_{168}O_{184}S$: 13570.44; observed: $[M + 8H]^{8+}$ *m/z* = 1698.30, $[M + 9H]^{9+}$ *m/z* = 1509.95, $[M + 10H]^{10+}$ *m/z* = 1359.00, $[M + 11H]^{11+}$ *m/z* = 1235.40, $[M + 12H]^{12+}$ *m/z* = 1132.60, $[M + 13H]^{13+}$ *m/z* = 1045.45.

L-Mb5-11 (L-35). Zn dust (40 mg) was added to peptide L-**34** (0.78 mg) in AcOH/H_2O (1:1; 0.78 mL), and the reaction mixture was incubated for 2 h at 37 °C. After Zn dust was removed by filtration, the mixture was purified by preparative HPLC (Cosmosil 5C18-AR300 column, a linear gradient of 26–56% CH_3CN in H_2O containing 0.05% (v/v) TFA over 90 min) to provide L-**35** (0.20 mg, 26% yield). MS(ESI) *m/z*, calcd for $C_{585}H_{860}N_{164}O_{176}S$: 13030.31; observed: $[M + 8H]^{8+}$ *m/z* = 1630.90, $[M + 9H]^{9+}$ *m/z* = 1449.80, $[M + 10H]^{10+}$ *m/z* = 1304.85, $[M + 11H]^{11+}$ *m/z* = 1186.30, $[M + 12H]^{12+}$ *m/z* = 1087.75, $[M + 13H]^{13+}$ *m/z* = 1003.90.

[Ala^{40,80}]-L-mMb5-11A (L-36a). Peptide L-**29a** (1.3 mg) was dissolved in desulfurization buffer [20 mM VA-044, 100 mM MESNa, 250 mM TCEP, 6 M guanidine, 100 mM phosphate (pH 6.5); 0.35 mL], and the reaction mixture was incubated for 2 h at 37 °C. The mixture was purified by preparative HPLC (Cosmosil 5C18-AR300 column, a linear gradient of 26–56% CH_3CN in H_2O containing 0.05% (v/v) TFA over 90 min) to provide L-**36a** (0.7 mg, 55%). MS(ESI) *m/z*, calcd for $C_{576}H_{849}N_{161}O_{175}S$: 12853.22; observed: $[M + 8H]^{8+}$ *m/z* = 1608.55, $[M + 9H]^{9+}$ *m/z* = 1429.85, $[M + 10H]^{10+}$ *m/z* = 1286.95, $[M + 11H]^{11+}$ *m/z* = 1170.05, $[M + 13H]^{13+}$ *m/z* = 990.15.

[Ala^{42,78}]-L-mMb5-11B (L-36b). By the same protocol for the synthesis of peptide L-**36a**, peptide L-**36b** was synthesized (0.5 mg, 45% yield) from peptide L-**29b** (1.1 mg). MS(ESI) *m/z*, calcd for $C_{583}H_{856}N_{164}O_{174}S$: 12970.29; observed: $[M + 8H]^{8+}$ *m/z* = 1623.10, $[M + 9H]^{9+}$ *m/z* = 1443.00, $[M + 10H]^{10+}$ *m/z* = 1298.70, $[M + 11H]^{11+}$ *m/z* = 1180.75, $[M + 12H]^{12+}$ *m/z* = 1082.45, $[M + 13H]^{13+}$ *m/z* = 999.20.

[Ala^{45,73}]-L-mMb5-11C (L-36c). By the same protocol for the synthesis of peptide L-**36a**, peptide L-**36c** was synthesized (0.5 mg, 42% yield) from peptide L-**29c** (1.2 mg). MS(ESI) *m/z*, calcd for $C_{581}H_{854}N_{164}O_{174}S$: 12944.27; observed: $[M + 8H]^{8+}$ *m/z* = 1619.90, $[M + 9H]^{9+}$ *m/z* = 1440.05, $[M + 10H]^{10+}$ *m/z* = 1296.05, $[M + 11H]^{11+}$ *m/z* = 1178.40, $[M + 13H]^{13+}$ *m/z* = 997.25.

[Cys(CH_2CONH_2)^{40,80}]-L-mMb5-11A (L-37a). Iodoacetamide in H_2O (0.5 M, 1.4 μL) was added to peptide L-**29a** (1.0 mg) in phosphate buffer [6 M guanidine, 200 mM phosphate (pH 7.3); 0.35 mL], and the reaction mixture was incubated for 1 h at room temperature. The mixture was purified by preparative HPLC (Cosmosil 5C18-AR300 column, a linear gradient of 25–55% CH_3CN in H_2O containing 0.05% (v/v) TFA over 90 min) to provide L-**37a** (0.4 mg, 36%). MS(ESI) *m/z*, calcd for $C_{580}H_{855}N_{163}O_{177}S_3$: 13031.21; observed: $[M + 8H]^{8+}$ *m/z* = 1631.30, $[M + 9H]^{9+}$ *m/z* = 1449.90, $[M + 10H]^{10+}$ *m/z* = 1305.05, $[M + 11H]^{11+}$ *m/z* = 1186.55, $[M + 12H]^{12+}$ *m/z* = 1087.60, $[M + 13H]^{13+}$ *m/z* = 1004.10.

[Cys(CH_2CONH_2)^{42,78}]-L-mMb5-11B (L-37b). By the same protocol for the synthesis of peptide L-**37a**, peptide L-**37b** was synthesized (0.6 mg, 69% yield) from peptide L-**29b** (0.9 mg). MS(ESI) *m/z*, calcd for $C_{587}H_{862}N_{166}O_{176}S_3$: 13148.28; observed: $[M + 8H]^{8+}$ *m/z* = 1645.50, $[M + 9H]^{9+}$ *m/z* = 1462.60, $[M + 10H]^{10+}$ *m/z* = 1316.55, $[M + 11H]^{11+}$ *m/z* = 1197.05, $[M + 12H]^{12+}$ *m/z* = 1097.25, $[M + 13H]^{13+}$ *m/z* = 1013.00.

[Cys(CH_2CONH_2)^{45,73}]-L-mMb5-11C (L-37c). By the same protocol for the synthesis of peptide L-**37a**, peptide L-**37c** was synthesized (0.6 mg, 48% yield) from peptide L-**29c** (1.2 mg). MS(ESI) *m/z*, calcd for $C_{585}H_{860}N_{166}O_{176}S_3$: 13122.26; observed: $[M + 8H]^{8+}$ *m/z* = 1642.05, $[M + 9H]^{9+}$ *m/z* = 1459.75, $[M + 10H]^{10+}$ *m/z* = 1314.20, $[M + 11H]^{11+}$ *m/z* = 1194.50, $[M + 13H]^{13+}$ *m/z* = 1010.90.

[Cys^{40}–Cys^{80}]-L-mMb5-11A (L-38a). 5,5'-Dithiobis(2-nitrobenzoic acid) in CH_3CN (1 mg/mL; 41 μL) was added to peptide L-**29a** (1.0 mg) in reaction buffer [6 M guanidine, 20 mM Tris (pH 8.0); 0.69 mL], and the reaction mixture was incubated for 30 min at room temperature. The mixture was purified by preparative HPLC (Cosmosil 5C4-AR300 column, a linear gradient of 24–54% CH_3CN in H_2O containing 0.05% (v/v) TFA over 90 min) to provide L-**38a** (0.3 mg, 30%). MS(ESI) *m/z*, calcd for $C_{576}H_{847}N_{161}O_{175}S_3$: 12915.15; observed: $[M + 7H]^{7+}$ *m/z* = 1847.55, $[M + 8H]^{8+}$ *m/z* = 1616.10, $[M + 9H]^{9+}$ *m/z* = 1436.80, $[M + 10H]^{10+}$ *m/z* = 1293.25.

[Cys^{42}–Cys^{78}]-L-mMb5-11B (L-38b). By the same protocol for the synthesis of peptide L-**38a**, peptide L-**38b** was synthesized (0.5 mg, 37% yield) from peptide L-**29b** (1.3 mg). MS(ESI) *m/z*, calcd for $C_{583}H_{854}N_{164}O_{174}S_3$: 13032.22; observed: [M

+ 7H]$^{7+}$ *m/z* = 1863.95, [M + 8H]$^{8+}$ *m/z* = 1630.95, [M + 9H]$^{9+}$ *m/z* = 1450.00, [M + 10H]$^{10+}$ *m/z* = 1304.85, [M + 11H]$^{11+}$ *m/z* = 1186.45.

[Cys^{45}–Cys^{73}]-L-mMb5-11C (L-38c). By the same protocol for the synthesis of peptide L-**38a**, peptide L-**38c** was synthesized (0.7 mg, 44% yield) from peptide L-**29c** (1.6 mg). MS(ESI) *m/z*, calcd for $C_{581}H_{852}N_{164}O_{174}S_3$: 13006.20; observed: [M + 7H]$^{7+}$ *m/z* = 1860.00, [M + 8H]$^{8+}$ *m/z* = 1627.40, [M + 9H]$^{9+}$ *m/z* = 1446.80, [M + 10H]$^{10+}$ *m/z* = 1302.20, [M + 12H]$^{12+}$ *m/z* = 1085.45.

[Cys^{40}–CH_2-CO-CH_2–Cys^{80}]-L-mMb5-11A (L-39a). 1,3-Dichloroacetone in CH_3CN (1 mg/mL; 18 μL) was added to peptide L-**29a** (1.4 mg) in reaction buffer [6 M guanidine, 20 mM Tris (pH 8.0); 0.93 mL], and the reaction mixture was incubated for 30 min at room temperature. The mixture was purified by preparative HPLC (Cosmosil 5C4-AR300 column, a linear gradient of 22–52% CH_3CN in H_2O containing 0.05% (v/v) TFA over 90 min) to provide L-**39a** (17% yield). The product was quantified by UV absorbance at 280 nm. MS(ESI) *m/z*, calcd for $C_{579}H_{851}N_{161}O_{176}S_3$: 12971.18; observed: [M + 7H]$^{7+}$ *m/z* = 1855.00, [M + 8H]$^{8+}$ *m/z* = 1623.25, [M + 9H]$^{9+}$ *m/z* = 1443.05, [M + 10H]$^{10+}$ *m/z* = 1298.70.

[Cys^{45}–CH_2-CO-CH_2–Cys^{73}]-L-mMb5-11C (L-39c). By the same protocol for the synthesis of peptide L-**39a**, peptide L-**39c** was synthesized (0.4 mg, 25% yield) from peptide L-**29c** (1.6 mg). MS(ESI) *m/z*, calcd for $C_{584}H_{856}N_{164}O_{175}S_3$: 13062.23; observed: [M + 7H]$^{7+}$ *m/z* = 1867.90, [M + 8H]$^{8+}$ *m/z* = 1634.95, [M + 9H]$^{9+}$ *m/z* = 1453.25.

[Cys^{45}–CH_2-CO-CH_2–Cys^{73}]-D-mMb5-11C (D-39c). By the identical procedure for the synthesis of peptide L-**39c**, peptide D-**39c** was synthesized (1.1 mg, 29% yield) from peptide D-**29c** (2.7 mg). MS(ESI) *m/z*, calcd for $C_{584}H_{856}N_{164}O_{175}S_3$: 13062.23; observed: [M + 7H]$^{7+}$ *m/z* = 1867.75, [M + 8H]$^{8+}$ *m/z* = 1634.35, [M + 9H]$^{9+}$ *m/z* = 1453.25, [M + 10H]$^{10+}$ *m/z* = 1307.85, [M + 11H]$^{11+}$ *m/z* = 1189.10.

[Cys^{40}–$(CH_2)_2$-SO_2-$(CH_2)_2$–Cys^{80}]-L-mMb5-11A (L-40a). Divinylsulfone in CH_3CN (1 μL/mL; 12 μL) was added to peptide L-**29a** (1.2 mg) in reaction buffer [6 M guanidine, 20 mM Tris (pH 8.0); 0.81 mL], and the reaction mixture was incubated for 30 min at room temperature. The mixture was purified by preparative HPLC (Cosmosil 5C4-AR300 column, a linear gradient of 22–52% CH_3CN in H_2O containing 0.05% (v/v) TFA over 90 min) to provide L-**40a** (10% yield). The product was quantified by UV absorbance at 280 nm. MS(ESI) *m/z*, calcd for $C_{580}H_{855}N_{161}O_{177}S_4$: 13035.17; observed: [M + 7H]$^{7+}$ *m/z* = 1863.85, [M + 8H]$^{8+}$ *m/z* = 1630.95, [M + 9H]$^{9+}$ *m/z* = 1449.75, [M + 10H]$^{10+}$ *m/z* = 1304.65.

[Cys^{45}–$(CH_2)_2$-SO_2-$(CH_2)_2$–Cys^{73}]-L-mMb5-11C (L-40c). By the same protocol for the synthesis of peptide L-**40a**, peptide L-**40c** was synthesized (0.4 mg, 37% yield) from peptide L-**29c** (1.2 mg). MS(ESI) *m/z*, calcd for $C_{585}H_{860}N_{164}O_{176}S_4$: 13126.23; observed: [M + 7H]$^{7+}$ *m/z* = 1877.15, [M + 8H]$^{8+}$ *m/z* = 1642.60, [M + 9H]$^{9+}$ *m/z* = 1460.00, [M + 10H]$^{10+}$ *m/z* = 1314.20, [M + 11H]$^{11+}$ *m/z* = 1194.70.

[Cys[40]–$(CH_2)_2$-SO_2-CH_2-SO_2-$(CH_2)_2$–Cys[80]]-L-mMb5-11A (L-41a). Bis(vinylsulfonyl)methane in CH_3CN (1 mg/mL; 19 μL) was added to peptide L-**29a** (1.0 mg) in reaction buffer [6 M guanidine, 20 mM Tris (pH 8.0); 0.65 mL], and the reaction mixture was incubated for 30 min at room temperature. The mixture was purified by preparative HPLC (Cosmosil 5C4-AR300 column, a linear gradient of 22–52% CH_3CN in H_2O containing 0.05% (v/v) TFA over 90 min) to provide L-**41a** (11% yield). The product was quantified by UV absorbance at 280 nm. MS(ESI) *m/z*, calcd for $C_{581}H_{857}N_{161}O_{179}S_5$: 13113.15; observed: $[M + 7H]^{7+}$ *m/z* = 1874.75, $[M + 8H]^{8+}$ *m/z* = 1640.65, $[M + 9H]^{9+}$ *m/z* = 1458.30, $[M + 10H]^{10+}$ *m/z* = 1312.80.

[Cys[45]–$(CH_2)_2$-SO_2-CH_2-SO_2-$(CH_2)_2$–Cys[73]]-L-mMb5-11C (L-41c). By the same protocol for the synthesis of peptide L-**41a**, peptide L-**41c** was synthesized (0.3 mg, 15% yield) from peptide L-**29c** (1.9 mg). MS(ESI) *m/z*, calcd for $C_{586}H_{862}N_{164}O_{178}S_5$: 13204.20; observed: $[M + 8H]^{8+}$ *m/z* = 1651.85, $[M + 9H]^{9+}$ *m/z* = 1468.45, $[M + 10H]^{10+}$ *m/z* = 1321.65, $[M + 11H]^{11+}$ *m/z* = 1201.45, $[M + 12H]^{12+}$ *m/z* = 1101.40, $[M + 13H]^{13+}$ *m/z* = 1016.80.

[Cys[45]–maleimide–Cys[73]]-L-mMb5-11C (L-42c). 3,4-Dibromomaleimide in CH_3CN (1 mg/mL; 36 μL) was added to peptide L-**29c** (1.4 mg) in reaction buffer [6 M guanidine, 20 mM Tris (pH 8.0); 0.94 mL], and the reaction mixture was incubated for 30 min at room temperature. The mixture was purified by preparative HPLC (Cosmosil 5C4-AR300 column, a linear gradient of 23–53% CH_3CN in H_2O containing 0.05% (v/v) TFA over 90 min) to provide L-**42c** (11% yield). The product was quantified by UV absorbance at 280 nm. MS(ESI) *m/z*, calcd for $C_{585}H_{853}N_{165}O_{176}S_3$: 13101.20; observed: $[M + 7H]^{7+}$ *m/z* = 1874.05, $[M + 8H]^{8+}$ *m/z* = 1639.80, $[M + 9H]^{9+}$ *m/z* = 1457.55, $[M + 10H]^{10+}$ *m/z* = 1311.60, $[M + 11H]^{11+}$ *m/z* = 1192.65.

Folding of Mb5-11 Variants (SH form). The lyophilized peptide was dissolved in guanidine solution [10 mM DTT, 6 M guanidine, 20 mM Tris (pH 8); 20 μM]. The denatured protein was subjected to repeated dialysis using Slide-A-Lyzer G2 dialysis cassette (cutoff 3.5 kDa, Thermo) against a 200-fold volume of dialysis buffer [50 mM acetate buffer (pH 4.5)] at 4 °C. For CD spectra measurement and thermal stability assay, the samples were subjected to dialysis to phosphate buffer [1 mM DTT, 50 mM phosphate (pH 7)]. For SPR analysis, the samples were subjected to dialysis to HEPES buffer [1 mM DTT, 20 mM HEPES (pH 7)]. The folded proteins were used for the following experiments without purification.

Folding of Mb5-11 Derivatives. The lyophilized peptide was dissolved in guanidine solution [6 M guanidine, 20 mM Tris (pH 8); 20 μM]. The denatured protein was subjected to repeated dialysis using Slide-A-Lyzer G2 dialysis cassette (cutoff 3.5 kDa, Thermo) against a 200-fold volume of dialysis buffer [50 mM acetate buffer (pH 4.5)] at 4 °C. For CD spectra measurement and thermal stability assay, the samples were subjected to dialysis to phosphate buffer [50 mM phosphate (pH 7)] by dialysis. For SPR analysis, the samples were subjected to dialysis to HEPES buffer [20 mM HEPES (pH 7)]. The folded proteins were used for the following experiments without purification.

Disulfide Bond Formation of Refolded Mb5-11 Variants. Refolded proteins L-**29a–c** (~ 20 μM) in acetate buffer [50 mM acetate (pH 4.5)] were dialyzed against a 200-fold volume of HEPES buffer [20 mM HEPES (pH 7.4)] and then oxidation buffer [1 mM cystamine, 20 mM HEPES (pH 7.4)] using Slide-A-Lyzer G2 dialysis cassette (cutoff 3.5 kDa, Thermo) at 4 °C. After oxidation for 42 h, the mixture was analyzed by LC–MS.

Oxime Ligation. Aniline hydrochloride in H_2O (1 M, 50 μL) and $MeONH_2$ in H_2O (20 mM, 23 μL) was added to the refolded protein D-**39c** (~ 12 μM, 0.40 mL) in acetate buffer [50 mM acetate (pH 4.5)], and the reaction mixture was incubated for 26 h at room temperature. The reaction was monitored by LC–MS analysis. The reaction mixture was dialyzed against a 200-fold volume of phosphate buffer [50 mM phosphate (pH 7)] or HEPES buffer [20 mM HEPES (pH 7)] four times at 4 °C. MS(ESI) *m/z*, calcd for $C_{585}H_{859}N_{165}O_{175}S_3$: 13091.26; observed: (former peak) $[M + 7H]^{7+}$ $m/z = 1871.20$, $[M + 8H]^{8+}$ $m/z = 1637.40$, $[M + 9H]^{9+}$ $m/z = 1455.90$, $[M + 10H]^{10+}$ $m/z = 1310.00$, $[M + 11H]^{11+}$ $m/z = 1191.30$. (latter peak) $[M + 7H]^{7+}$ $m/z = 1871.30$, $[M + 8H]^{8+}$ $m/z = 1637.40$, $[M + 9H]^{9+}$ $m/z = 1455.65$, $[M + 10H]^{10+}$ $m/z = 1310.30$, $[M + 11H]^{11+}$ $m/z = 1191.30$.

CD Spectra Measurement. Proteins (~ 10 μM) in phosphate buffer [50 mM phosphate buffer (pH 7)] were used for measurement. CD spectra were recorded on a J-1500 circular dichroism spectrometer (JASCO) at 25 °C. The thermal denaturation was monitored by observing the transition of CD signal at 203 nm during heating at a rate of 1 °C/min (protein concentration: ~ 10 μM). T_m values were defined as temperatures at the time points when half of the protein molecules are unfolded. T_m values were determined from single assays.

SPR Analysis. Binding of synthetic monobodies to L-MCP-1 or D-MCP-1 was analyzed using a Biacore X-100 systems (Cytiva). Biotinylated L-MCP-1 or D-MCP-1 [31] was immobilized on an SA sensor chip (Cytiva) with HBS-EP$^+$ buffer (Cytiva) as a running buffer (~ 100 RU). The analyte protein solution was injected at a rate of 30 μL/min into the flow system in HBS-EP$^+$ buffer. Association of all the analytes was evaluated for 90 s, then dissociation was monitored for 600 s. Glycine 2.0 solution (Cytiva) was used for regeneration. Binding affinities were determined from triplicate assays by fitting a 1:1 binding model.

References

1. Koide S et al (1998) The fibronectin type III domain as a scaffold for novel binding proteins. J Mol Biol 284:1141–1151
2. Koide A et al (2012) Teaching an old scaffold new tricks: Monobodies constructed using alternative surfaces of the FN3 scaffold. J Mol Biol 415:393–405

3. Hantschel O et al (2020) Monobodies as enabling tools for structural and mechanistic biology. Curr Opin Struct Biol 60:167–174
4. Parker MH et al (2005) Antibody mimics based on human fibronectin type three domain engineered for thermostability and high-affinity binding to vascular endothelial growth factor receptor two. Protein Eng Des Sel 18:435–444
5. Mitchell T et al (2014) Pharmacologic profile of the adnectin BMS-962476, a small protein biologic alternative to PCSK9 antibodies for low-density lipoprotein lowering. J Pharmacol Exp Ther 350:412–424
6. Gebauer M, Skerra A (2020) Engineered protein scaffolds as next generation therapeutics. Annu Rev Pharmacol Toxicol 60:391–415
7. Vaisman-Mentesh A et al (2020) The molecular mechanisms that underlie the immune biology of anti-drug antibody formation following treatment with monoclonal antibodies. Front Immunol 11:1951–1966
8. Mamluk R et al (2010) Anti-tumor effect of CT-322 as an adnectin inhibitor of vascular endothelial growth factor receptor-2. mAbs 2:199–208
9. Tolcher AW et al (2011) Phase I and pharmacokinetic study of CT-322 (BMS-844203), a targeted adnectin inhibitor of VEGFR-2 based on a domain of human fibronectin. Clin Cancer Res 17:363–371
10. Schmidt N et al (2024) Development of mirror-image monobodies targeting the oncogenic BCR::ABL1 kinase. Nat Commun 15:10724
11. Dawson PE et al (1994) Synthesis of proteins by native chemical ligation. Science 266:776–779
12. Goldberg SD et al (2016) Engineering a targeted delivery platform using Centyrins. Protein Eng Des Sel 29:563–572
13. Hackeng TM et al (1999) Protein synthesis by native chemical ligation: Expanded scope by using straightforward methodology. Proc Natl Acad Sci USA 96:10068–10073
14. Blanco-Canosa JB et al (2015) Chemical protein synthesis using a second-generation *N*-acylurea linker for the preparation of peptide-thioester precursors. J Am Chem Soc 137:7097–7209
15. Blanco-Canosa JB, Dawson PE (2008) An efficient Fmoc-SPPS approach for the generation of thioester peptide precursors for use in native chemical ligation. Angew Chem Int Ed 47:6851–6855
16. Sakamoto K et al (2017) 1,2,4-Triazole-aided native chemical ligation between peptide-*N*-acyl-*N*′-methyl-benzimidazolinone and cysteinyl peptide. Chem Commun 53:12236–12239
17. Wang JX et al (2015) Peptide *o*-aminoanilides as crypto-thioesters for protein chemical synthesis. Angew Chem Int Ed 54:2194–2198
18. Jin K, Li X (2018) Advances in native chemical ligation–desulfurization: A powerful strategy for peptide and protein synthesis. Chem Eur J 24:17397–17404
19. Tsuda S et al (2018) Easy-to-attach/detach solubilizing-tag-aided chemical synthesis of an aggregative capsid protein. Angew Chem Int Ed 57:2105–2109
20. Plaxco KW et al (1996) Rapid refolding for a proline-rich all-β-sheet fibronectin type III module. Proc Natl Acad Sci USA 93:10703–10706
21. Clarke J et al (1997) Folding and stability of a fibronectin type III domain of human tenascin. J Mol Biol 270:771–778
22. Koide A et al (2001) Stabilization of a fibronectin type III domain by the removal of unfavorable electrostatic interaction on the protein surface. Biochemistry 40:10326–10333
23. Dintzis HM et al (1993) A comparison of the immunogenicity of a pair of enantiomeric proteins. Proteins Struct Funct Genet 16:306–308
24. Uppalapati M et al (2016) A potent D-protein antagonist of VEGF-A is nonimmunogenic, metabolically stable, and longer-circulating in vivo. ACS Chem Biol 11:1058–1065
25. Marinec PS et al (2021) A non-immunogenic bivalent D-protein potently inhibits retinal vascularization and tumor growth. ACS Chem Biol 16:548–556
26. Gongora-Beńítez M et al (2014) Multifaced roles of disulfide bonds. Peptide as therapeutics. Chem Rev 114:901–926

27. Gilbreth RN et al (2014) Stabilization of the third fibronectin type III domain of human tenascin-C through minimal mutation and rational design. Protein Eng Des Sel 27:411–418
28. Maruyama K et al (1999) 2,2'-Bispyridyl disulfide rapidly induces intramolecular disulfide bonds in peptides. Peptides 20:881–884
29. Hughes CE, Nibbs RJB (2018) A guide to chemokines and their receptors. FEBS J 285:2944–2971
30. Xu M et al (2021) Roles of the CCL2–CCR2 signaling axis in cancer: Mechanisms and therapeutic targeting. Cell Prolif 54:e13115
31. Hayashi G et al (2024) Generating a mirror-image monobody targeting MCP-1 via TRAP display and chemical protein synthesis. Nat Commun 15:10723
32. Ishizawa T et al (2013) TRAP display: A high-speed selection method for the generation of functional polypeptides. J Am Chem Soc 135:5433–5440
33. Fang GM et al (2011) Protein chemical synthesis by ligation of peptide hydrazides. Angew Chem Int Ed 50:7645–7649
34. Tian X et al (2016) Optimal peptide hydrazide ligation with C-terminus Asp, Asn, and Gln hydrazides. Tetrahedron Lett 57:4264–4267
35. Flood DT et al (2018) Leveraging the Knorr pyrazole synthesis for the facile generation of thioester surrogates for use in native chemical ligation. Angew Chem Int Ed 57:11634–11639
36. Asahina Y et al (2015) Chemical synthesis of O-glycosylated human interleukin-2 by the reverse polarity protection strategy. Angew Chem Int Ed 54:8226–8230
37. Veber DF et al (1977) Isonicotinyloxycarbonyl, a novel amino protecting group for peptide synthesis. J Org Chem 42:3286–3288
38. Chakraborty K et al (2005) Protein stabilization by introduction of cross-strand disulfides. Biochemistry 44:14638–14646
39. Gunnoo SB, Madder A (2016) Chemical protein modification through cysteine. ChemBioChem 17:529–553
40. Renault K et al (2018) Covalent modification of biomolecules through maleimide-based labeling strategies. Bioconjug Chem 29:2497–2513
41. Goel N, Stephens S (2010) Certolizumab pegol. mAbs 2:137–147
42. Ellman GL (1959) Tissue sulfhydryl groups. Arch Biochem Biolys 82:70–77
43. Fairlie DP, de Araujo AD (2016) Stapling peptides using cysteine crosslinking. Biopolymers 106:843–852
44. Zaretsky S, Yudin AK (2017) Recent advances in the synthesis of cyclic pseudopeptides. Drug Discov Today Technol 26:3–10
45. Skowron KJ et al (2019) Recent structural advances in constrained helical peptides. Med Res Rev 39:749–770
46. Assem N et al (2015) Acetone-linked peptides: A convergent approach for peptide macrocyclization and labeling. Angew Chem Int Ed 54:8665–8668
47. Chua K et al (2015) Small cyclic agonists of iron regulatory hormone hepcidin. Bioorg Med Chem Lett 25:4961–4969
48. Behrens CR et al (2015) Antibody–drug conjugates (ADCs) derived from interchain cysteine cross-linking demonstrate improved homogeneity and other pharmacological properties over conventional heterogeneous ADCs. Mol Pharmaceutics 12:3986–3998
49. White AM et al (2022) Late-stage functionalization with cysteine staples generates potent and selective melanocortin receptor-1 agonists. J Med Chem 65:12956–12969
50. Dirksen A et al (2006) Nucleophilic catalysis of oxime ligation. Angew Chem Int Ed 45:7581–7584
51. Asahina Y et al (2013) Chemoenzymatic synthesis of the immunoglobulin domain of Tim-3 carrying a complex-type N-glycan by using a one-pot ligation. Angew Chem Int Ed 52:9733–9737

Chapter 5
Conclusions

Abstract The author established synthetic protocols for two protein domains derived from the T cell immunoreceptor with immunoglobulin and immunoreceptor tyrosine-based inhibitory motif domain (TIGIT) and human serum albumin (HSA). Additionally, the author developed a synthetic variant of mirror-image monobody, an artificial protein scaffold for specific target binding. These protein domains are valuable tools for developing less-immunogenic protein therapeutics.

Keywords Protein domain · Mirror-image protein · Chemical protein synthesis · Monobody · HSA · TIGIT

The author established a protocol to synthesize the extracellular domain of the T cell immunoreceptor with immunoglobulin and immunoreceptor tyrosine-based inhibitory motif domain ($TIGIT^{22-141}$). During the synthesis, a trityl-type solubilizing auxiliary Trt-K_{10} was employed to improve the solubility of the synthetic intermediates as well as to temporarily protect the sulfhydryl groups of cysteine residues in the native sequence. The resulting synthetic $TIGIT^{22-141}$ showed reasonable binding to the target CD155 protein.

The author investigated the chemical synthesis of human serum albumin domain III ($HSA^{380-585}$). The sequence was constructed by convergent native chemical ligations from seven peptide segments. The symmetric structures and biological functions of synthetic L-$HSA^{380-585}$ and D-$HSA^{380-585}$ were confirmed by analyzing the circular dichroism spectra and site II-dependent molecular interactions.

The author established straightforward synthetic processes for monobody variants with cystine substitutions. The variants maintained sufficient target binding capability. The D-enantiomer of a monobody variant induced less anti-drug antibody generation compared with the L-enantiomer. The structure–activity relationship study of monobody variants with different substitution patterns and cysteine modifications led to identification of a monobody scaffold with potent target binding affinity and high thermal stability.

In summary, the author conducted chemical syntheses and functional evaluations of several protein domains of pharmaceutical importance. The use of these synthetic

N. Iwamoto, *Synthesis of Functional Characterization of Protein Domains for the Development of Mirror-Image Therapeutics*, Springer Theses,
https://doi.org/10.1007/978-981-95-5366-2_5

protein domains should help the future development of practical D-protein therapeutics through improving the pharmacokinetic profiles and facilitating the synthetic access.

5.1 Curriculum Vitae

Naoya Iwamoto

Laboratory of Medicinal Chemistry, Kyoto Pharmaceutical University

Yamashina-ku, Kyoto

607-8412, Japan

e-mail: niwamoto@mb.kyoto-phu.ac.jp.

Web: https://researchmap.jp/naoyaiwamoto

Appointments

- Assistant Professor, Kyoto Pharmaceutical University (Oct. 2024).

Education

- PhD in Pharmaceutical Sciences, Kyoto University (2021–2024), advisor Prof. Hiroaki Ohno and Prof. Shinya Oishi
- MS in Pharmaceutical Sciences, Kyoto University (2019–2021), advisor Prof. Hiroaki Ohno and Prof. Shinya Oishi
- Bachelor of Pharmaceutical Sciences, Kyoto University (2015–2019).

Awards

- SUNBOR Scholarship (2024)
- JSPS Research Fellowships for Young Scientists (2022–2024)
- Fujita Memorial Scholarship (2021–2022).

Research Statement and Interests

My research aims to explore unique molecular properties of therapeutic proteins through chemical synthesis approach. I am particularly interested in incorporating non-natural structural units into proteins and evaluate their effects on the functions of the parent proteins.

MIX
Papier aus verantwortungsvollen Quellen
Paper from responsible sources
FSC® C105338

If you have any concerns about our products,
you can contact us on
ProductSafety@springernature.com

In case Publisher is established outside the EU,
the EU authorized representative is:
Springer Nature Customer Service Center GmbH
Europaplatz 3, 69115 Heidelberg, Germany

Printed by Libri Plureos GmbH
in Hamburg, Germany